An Introduction
to Astrophysical
Fluid Dynamics

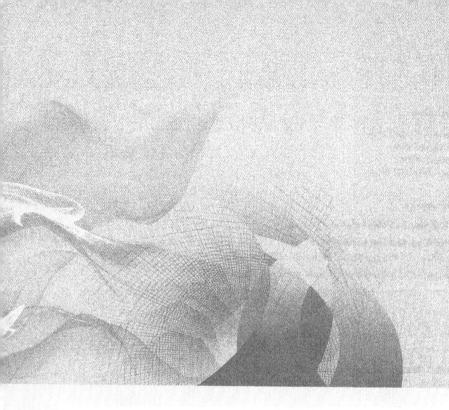

An Introduction to Astrophysical Fluid Dynamics

Michael J Thompson

University of Sheffield, UK

Imperial College Press

Published by

Imperial College Press
57 Shelton Street
Covent Garden
London WC2H 9HE

Distributed by

World Scientific Publishing Co. Pte. Ltd.
5 Toh Tuck Link, Singapore 596224
USA office: 27 Warren Street, Suite 401-402, Hackensack, NJ 07601
UK office: 57 Shelton Street, Covent Garden, London WC2H 9HE

British Library Cataloguing-in-Publication Data
A catalogue record for this book is available from the British Library.

ISBN   1-86094-615-1
ISBN   1-86094-633-X (pbk)

Printed by Mainland Press Pte Ltd

Preface

Fluid dynamical processes play a central role in almost all areas of astrophysics. This role is not always acknowledged in undergraduate courses or even graduate courses in the field, however.

This book could reasonably form the basis for a one-semester graduate course in astrophysical fluid dynamics. Over a number of years the material upon which the book is based has been given in lecture courses to graduate students (PhD and Masters-level), and also occasionally to advanced-level undergraduate students, studying astrophysics, physics or applied mathematics in the University of London. The scope of the material has been expanded somewhat for the present book, in particular giving an introduction to concepts of numerical computations in astrophysical fluid dynamics.

The treatment starts from a continuum description of a fluid (be it gas or liquid) and establishes the equations of fluid dynamics, so that no prior study of fluid dynamics is required. After introducing physical concepts necessary for the application to astrophysics, a number of different aspects and application areas are presented. The emphasis is mainly on the fluid dynamical properties rather than, say, radiative transfer or particle dynamics. While the choice of applications inevitably reflects to some extent my own research interests in solar and stellar physics, this introduction is intended to provide readers with the foundations of astrophysical fluid dynamics from which to study applications in areas in addition to the ones developed further here. There are some excellent advanced texts — I would mention in particular Shu's two-volume *The Physics of Astrophysics* (*Volume I: Radiation* and *Volume II: Gas Dynamics*) and Mihalas & Mihalas's *Foundations of Radiation Hydrodynamics* — for the advanced graduate student and researcher, to which it is hoped that the present volume will provide a useful intermediate step.

I am grateful to Richard Nelson and Jørgen Christensen-Dalsgaard for comments on parts of an earlier draft of the text and for discussions. I am also happy to acknowledge the heritage of material in the original lecture courses from earlier courses given by Jørgen Christensen-Dalsgaard, Søren Frandsen, Douglas Gough and John Papaloizou. I thank those teachers who introduced me to this fascinating subject, particularly John Green and Douglas Gough. Finally, I thank Kate for her patience and constant support.

M. J. Thompson
September 2004

Contents

Chapter 1

Basic Fluid Equations

Fluid dynamics is the continuum description of the flow of a large number of particles. Such a description is widely applicable in astrophysical problems, and fluid dynamical processes play a key role in many areas of astrophysics. In this book the fluid under consideration will generally be a gas, though the equations of fluid dynamics can also be applied to describing the motion of a collection of stars, or even galaxies, provided that one is interested in the collective behaviour on sufficiently large scales.

The term *fluid* in general refers to gases and liquids. Fluids are distinguished from solids in that solids have rigidity. Both solids and fluids deform when a stress is applied to them; but, unlike a solid, a simple fluid has no tendency to return to its original state when the applied stress is removed.

The continuum description is fundamental to the fluid approach to describing the dynamics of a collection of particles. The domain of validity of the continuum description is determined by comparing the collisional mean free path l of the particles with the macroscopic length scale L of interest in the problem. If $l \ll L$ then it is reasonable to introduce the concept of a fluid volume element, whose linear size is much larger than l but much smaller than L. The number of particles inside a fluid element is large, and we can associate with the fluid element a bulk velocity $\boldsymbol{u}$. Individual particle velocities have a random component in addition to $\boldsymbol{u}$ but, because the mean free path is small, the random motion does not immediately take the particle far from its neighbouring particles because the particle travels only a distance of order l before undergoing a collision and changing its direction. We can also associate with the fluid element other macroscopic properties such as a density ρ (total mass of the particles inside the element, divided by its volume). Over a short time interval from time t to time $t + \delta t$

1

we may define the fluid element to transform by translating each point of the element by an amount $u(r,t)\delta t$, where $u(r,t)$ is the local mean velocity at the position r of the element. By virtue of the above considerations, the fluid element will still contain essentially the same number of particles at $t + \delta t$ as it did at time t, and moreover they will be almost all the same particles as before. Hence the macroscopic properties of the fluid element will evolve only slowly, and by a diffusive process.

For further discussion of the continuum description and the fluid approach, see e.g. Batchelor (1967) and Shu (1992).

If the mean free path of the particles is not much smaller than the macroscopic scale of interest, then the appropriate description of the collective properties of the particles is kinetic theory. The equations of fluid dynamics can indeed be derived from the microscopic basis of kinetic theory. For a presentation of this approach, see Shu (1992). Here we shall instead assume the continuum description from the outset and see how simple considerations of the motion of the fluid, and the forces acting on it, lead to the fluid dynamical equations.

1.1   The Material Derivative

The fluid properties, such as its density ρ and velocity u, will in general be functions of position r and of time t. We shall always use $\partial/\partial t$ to denote the rate of change of some quantity with respect to time *at a fixed position in space*. In describing fluids it is also very useful to define the *material derivative*, which will be denoted D/Dt: this is the rate of change of some quantity with respect to time but travelling along with the fluid.

Let $f(r,t)$ be any quantity, for example, temperature of the fluid. It may happen that the temperature of all individual parcels of fluid is not changing with time, so the material derivative Df/Dt is zero; but if some fluid is hotter than other fluid then the temperature at a fixed point in space may still change with time as fluid of different temperature passes the point at which the temperature is measured. In fact, in that case, $\partial f/\partial t = -u \cdot \nabla f$ where $u(r,t)$ is the velocity of the fluid. More generally, the material derivative is related to the rate of change at a fixed point in space as

$$\frac{Df}{Dt} = \frac{\partial f}{\partial t} + u \cdot \nabla f . \tag{1.1}$$

Equation (1.1) can be derived by considering the change in f when following the fluid, over a short period of time δt, in the limit as δt tends to zero. Since (correct to first order in δt) the fluid element will have moved from r at time t to $r + u\delta t$ at time $t + \delta t$, the material derivative is

$$\frac{Df}{Dt} = \lim_{\delta t \to 0} \left(\frac{f(r + u\,\delta t, t + \delta t) - f(r,t)}{\delta t} \right)$$

and Eq. (1.1) follows.

1.2 The Continuity Equation

Consider a volume V, which is fixed in space, enclosed by a surface S on which n is the outward-pointing normal vector (Fig. 1.1). The total mass of fluid in V is $\int_V \rho dV$, where $\rho(r,t)$ is the density of the fluid. The time derivative of the mass in V is the mass flux into V across its surface S, i.e.

$$\frac{\mathrm{d}}{\mathrm{d}t} \int_V \rho\,\mathrm{d}V = - \int_S (\rho u) \cdot n\,\mathrm{d}S. \tag{1.2}$$

Since V is a volume fixed in space, the time derivative on the left of Eq. (1.2) can be taken inside the integral and becomes a derivative at fixed position in space. The surface term on the right-hand side of the equation can be re-expressed as a volume integral using the divergence theorem. Hence we obtain

$$\int_V \frac{\partial \rho}{\partial t}\,\mathrm{d}V = - \int_V \nabla \cdot (\rho u)\,\mathrm{d}V.$$

Since this holds for any arbitrary volume V in the fluid, it follows that

$$\frac{\partial \rho}{\partial t} + \nabla \cdot (\rho u) = 0. \tag{1.3}$$

This is the *continuity equation* (or mass conservation equation). Combining Eqs. (1.1) and (1.3) the continuity equation can also be expressed as

$$\frac{D\rho}{Dt} + \rho \nabla \cdot u = 0. \tag{1.4}$$

1.3 The Momentum Equation

One can similarly derive a momentum equation, or equation of motion, for the fluid by considering the rate of change of the total momentum of the fluid inside a volume V. It turns out to be easiest to consider a volume

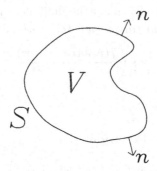

Fig. 1.1   An arbitrary volume of fluid V, with surface S and outward-pointing normal $\boldsymbol{n}$.

moving with the fluid, so that no fluid is flowing across its surface into or out of V. The momentum of the fluid in V is $\int_V \rho \boldsymbol{u}\, dV$, and the rate of change of this momentum is equal to the net force acting on the fluid in volume V. These are of two kinds. First there are body forces, such as gravity, which act on the particles inside V: their net effect is a force

$$\int_V \rho \boldsymbol{f}\, dV$$

where $\boldsymbol{f}$ is the body force per unit mass. (Note that force per unit mass has dimensions of acceleration.) For example, $\boldsymbol{f}$ could be the gravitational acceleration $\boldsymbol{g}$. The second kind of forces acting are surface forces – forces exerted on the surface S of V by the surrounding fluid. In an *inviscid* fluid, such as we shall mostly be considering, the surface force acts normally to the surface and its net effect is

$$\int_S -p\boldsymbol{n}\, dS \,,$$

p being the pressure. There is no flux of momentum across the surface carried by fluid parcels moving, since by definition none crosses the surface of a material volume. Equating force to change of momentum we obtain

$$\frac{d}{dt} \int_V \rho \boldsymbol{u}\, dV \;=\; \int_S -p\boldsymbol{n}\, dS \;+\; \int_V \rho \boldsymbol{f}\, dV \,. \qquad (1.5)$$

Since V is a material volume, when the time derivative is taken inside the integral it becomes a material derivative; but the product ρdV is the mass of a fluid element and is invariant following the motion so its material derivative is zero. Thus

$$\frac{\mathrm{d}}{\mathrm{d}t} \int_V \rho \boldsymbol{u} \, \mathrm{d}V \;=\; \int_V \rho \frac{D\boldsymbol{u}}{Dt} \, \mathrm{d}V \tag{1.6}$$

and hence, applying the divergence theorem to the surface integral in Eq. (1.5), we obtain

$$\int_V \rho \frac{D\boldsymbol{u}}{Dt} \, \mathrm{d}V \;=\; \int_V (-\nabla p + \rho \boldsymbol{f}) \, \mathrm{d}V \;.$$

Since this holds for any arbitary material volume V, it follows that

$$\rho \frac{D\boldsymbol{u}}{Dt} \;\equiv\; \rho \left(\frac{\partial \boldsymbol{u}}{\partial t} + (\boldsymbol{u} \cdot \nabla)\boldsymbol{u} \right) \;=\; (-\nabla p + \rho \boldsymbol{f}) \;. \tag{1.7}$$

This is the momentum equation for an inviscid fluid.

In a general viscous fluid (it doesn't need to be as extreme an example as treacle!) the ith component of the force exerted on surface S by the surrounding fluid is not just $\int_S -p\,n_i \, \mathrm{d}S$ but is $\int_S \sigma_{ij} n_j \, \mathrm{d}S$, where σ_{ij} is the *stress tensor*. (Note here that the summation convention is used, so that if an index is repeated it should be summed over. A non-repeated index denotes a component of a vector or tensor. See Appendix A. Also note that throughout this book we shall use $\boldsymbol{r}$ or $\boldsymbol{x}$ to denote vector position; but for its ith component form we always write x_i.) For gases and simple liquids it is found that

$$\sigma_{ij} \;=\; -p\delta_{ij} + \mu \left(\frac{\partial u_i}{\partial x_j} + \frac{\partial u_j}{\partial x_i} - \frac{2}{3}(\nabla \cdot \boldsymbol{u})\delta_{ij} \right) \tag{1.8}$$

where μ is the so-called dynamical viscosity: see e.g. Batchelor (1967), Landau & Lifshitz (1959). Also δ_{ij} is the Kronecker delta: see Appendix A. Now

$$\int_S \sigma_{ij} \, n_j \, \mathrm{d}S \;=\; \int_V \frac{\partial}{\partial x_j} \sigma_{ij} \, \mathrm{d}V$$

(divergence theorem), and so if μ is a constant it follows that the equation of motion for a viscous fluid is

$$\rho \frac{D\boldsymbol{u}}{Dt} \;=\; -\nabla p + \rho \boldsymbol{f} + \mu \left(\nabla^2 \boldsymbol{u} + \frac{1}{3}\nabla(\nabla \cdot \boldsymbol{u}) \right) \;. \tag{1.9}$$

With viscosity included, Eq. (1.9) is called the Navier-Stokes equation. Its inviscid form, Eq. (1.7), is called the Euler equation. Throughout most of this book we shall neglect viscosity: the justification of this approximation in astrophysical contexts will be seen in Section 2.8. However, viscosity plays a key role in some applications, notably in astrophysical accretion disks which we discuss in Chapter 9.

1.4 Newtonian Gravity

A mass m' at position r' exerts on any other mass m at position r an attractive force proportional to the product of the two masses and inversely proportional to the square of the distance between them, directed towards mass m':

$$\boldsymbol{F} \;=\; m\boldsymbol{g}(\boldsymbol{r}) \;\equiv\; -\,\frac{Gm\,m'}{|\boldsymbol{r}-\boldsymbol{r}'|^2}\,\frac{(\boldsymbol{r}-\boldsymbol{r}')}{|\boldsymbol{r}-\boldsymbol{r}'|} \;\equiv\; -\,\frac{Gm\,m'(\boldsymbol{r}-\boldsymbol{r}')}{|\boldsymbol{r}-\boldsymbol{r}'|^3}\,. \tag{1.10}$$

Note that $(\boldsymbol{r}-\boldsymbol{r}')/|\boldsymbol{r}-\boldsymbol{r}'|$ is the unit vector along the line of action of the force. Now

$$\nabla\!\left(\frac{1}{|\boldsymbol{r}-\boldsymbol{r}'|}\right) \;=\; \frac{-(\boldsymbol{r}-\boldsymbol{r}')}{|\boldsymbol{r}-\boldsymbol{r}'|^3} \tag{1.11}$$

(the derivatives are with respect to $\boldsymbol{r}$: they treat $\boldsymbol{r}'$ as a constant vector), so the gravitational acceleration $\boldsymbol{g}(\boldsymbol{r})$ can be written as the gradient of a potential function $\psi(\boldsymbol{r})$:

$$\boldsymbol{g} \;=\; -\nabla\psi\,, \quad \text{where} \quad \psi \;=\; \frac{-Gm'}{|\boldsymbol{r}-\boldsymbol{r}'|}\,. \tag{1.12}$$

Similarly, the gravitational field due to a fluid can be written as a potential, namely the sum of the potentials due to all the fluid elements. The mass of a fluid element of volume $\mathrm{d}V'$ at position $\boldsymbol{r}'$ is $\rho(\boldsymbol{r}')\mathrm{d}V'$, so the total gravitational potential is

$$\psi(\boldsymbol{r}) \;=\; \int_{V'} \frac{-G\rho(\boldsymbol{r}')}{|\boldsymbol{r}-\boldsymbol{r}'|}\,\mathrm{d}V'\,, \tag{1.13}$$

where the integration is over the whole volume of the fluid. The gravitational acceleration is $-\nabla\psi$.

Using the result

$$\nabla^2\!\left(\frac{1}{|\boldsymbol{r}-\boldsymbol{r}'|}\right) \;=\; -4\pi\,\delta(\boldsymbol{r}-\boldsymbol{r}') \tag{1.14}$$

(δ being the Dirac delta function in 3-D space), Eq. (1.13) can be rewritten as a partial differential equation, *Poisson's equation*:

$$\nabla^2 \psi = 4\pi G \rho . \tag{1.15}$$

1.5 The Mechanical and Thermal Energy Equations

If one takes Newton's third law, $F = ma = m(dv/dt)$ and multiplies by velocity v, one obtains that rate of work of the forces, Fv, is equal to the rate of change of kinetic energy, $d(\frac{1}{2}mv^2)/dt$. Similarly, taking the dot product of the equation of motion for a fluid, (1.7), with the fluid velocity $\boldsymbol{u}$ yields

$$\frac{D}{Dt}\left(\frac{1}{2}\boldsymbol{u}^2\right) = -\frac{1}{\rho}\boldsymbol{u}\cdot\nabla p + \boldsymbol{u}\cdot\boldsymbol{f} . \tag{1.16}$$

Equation (1.16) says that the rate of change of the kinetic energy of a unit mass of fluid is equal to the rate at which work is done on the fluid by pressure and body forces. This is sometimes called the mechanical energy equation.

An equation for the total energy — kinetic and internal thermal energy — can be derived in the same manner as was the momentum equation in Section 1.3. Let the internal energy per unit mass of fluid be U. Then the rate of change of kinetic plus internal energy of a material volume (i.e. one moving with the fluid) must be equal to the rate of work done on the fluid by surface and body forces, plus the rate at which heat is added to the fluid. Heat can be added in two ways: one is by its being generated at a rate ϵ per unit mass within the fluid volume (e.g. by nuclear reactions), while the second is by the heat flux $\boldsymbol{F}$ across the surface S (e.g. radiative heat flux). Thus

$$\frac{d}{dt}\int_V \left(\frac{1}{2}\boldsymbol{u}^2 + U\right)\rho\,dV \tag{1.17}$$

$$= \int_S \boldsymbol{u}\cdot(-p\boldsymbol{n})\,dS + \int_V \boldsymbol{u}\cdot\boldsymbol{f}\rho\,dV + \int_V \epsilon\rho dV - \int_S \boldsymbol{F}\cdot\boldsymbol{n}dS .$$

In the same way as for the momentum equation, one rewrites all the surface integrals in this equation as volume integrals, using the divergence theorem. The resulting equation holds for an arbitrary volume V and so one deduces that

$$\rho \left(\frac{D}{Dt} \left(\frac{1}{2} u^2 \right) + \frac{DU}{Dt} \right) = -\nabla \cdot (p u) + \rho u \cdot f + \rho \epsilon - \nabla \cdot F . \quad (1.18)$$

One can derive an equation for the thermal energy alone by dividing Eq. (1.18) by the density and then subtracting the kinetic energy equation (1.16) to obtain

$$\frac{DU}{Dt} = \frac{p}{\rho^2} \frac{D\rho}{Dt} + \epsilon - \frac{1}{\rho} \nabla \cdot F . \quad (1.19)$$

The divergence of u has been replaced by $-\rho^{-1} D\rho/Dt$ using the continuity equation (1.4).

Noting that the volume per unit mass is just the reciprocal of the density, i.e. $V = \rho^{-1}$, we recognise the thermal energy equation (1.19) as a statement of the first law of thermodynamics:

$$dU = (-p)dV + \delta Q, \quad (1.20)$$

that is, the change in the internal energy is equal to the work $(-p)dV$ done (on the fluid) plus the heat added. Note that V, U, p are properties of the fluid (in fact they are thermodynamic state variables) and we denote changes in them with the symbol "d". In contrast, there is no such property as the heat content and so we cannot speak of the change of heat content. Instead, we can only speak of the heat added, and we therefore use a different notation, i.e. δQ.

Equations (1.16)-(1.19) can be generalized to include a viscous stress term. In that case, $-\rho^{-1} u \cdot \nabla p$ in Eq. (1.16) and $u \cdot (-pn)$ in equation (1.18) are replaced by $\rho^{-1} u_i \partial \sigma_{ij}/\partial x_j$ and $u_i \sigma_{ij} n_j$ respectively, where σ_{ij} is the stress tensor as in Eq. (1.8). The consequence for the thermal energy equation (1.19) is that kinetic energy is converted to heat by viscosity, so that one obtains an additional heating term similar to ϵ. This is discussed in more detail in Chapter 9.

With some effort, one can use the above equations to derive an integral equation (sometimes also referred to as the total energy equation) for the rate of change of the total energy (kinetic plus internal plus gravitational potential energy) for the whole fluid volume:

$$\frac{d}{dt} \int_V \left(\frac{1}{2} u^2 + U + \frac{1}{2} \psi \right) \rho dV + \int_V \nabla \cdot \left[\left(\frac{1}{2} u^2 + U + \frac{p}{\rho} + \psi \right) \rho u \right] dV$$

$$= \int_V (\rho \epsilon - \nabla \cdot F) dV , \quad (1.21)$$

where V is a fixed volume enclosing the whole fluid: e.g. Cox (1980). In deriving the above equation it is helpful first to establish from eq. (1.13) that $\int_V (\partial \rho / \partial t) \psi \mathrm{d}V = \int_V (\rho \partial \psi / \partial t) \mathrm{d}V$ where V is the whole region occupied by the fluid.

The volume integrals of divergence terms in Eq. (1.21) can of course be re-expressed as surface integrals. If the flux in square brackets in the second term on the left of Eq. (1.21) vanishes at the surface of V, which might for example represent the interior of a star, then the total energy in V can only change through internal heat sources/sinks (ϵ) or heat flux ($\boldsymbol{F}$) across the surface.

1.6 A Little More Thermodynamics

The second law of thermodynamics states that

$$\delta Q = T \, \mathrm{d}S \,, \tag{1.22}$$

where S is a thermodynamic state variable, the *specific entropy* (i.e. the entropy per unit mass). Combining this with the first law, Eq. (1.20), yields

$$\mathrm{d}U = T \mathrm{d}S - p \mathrm{d}V \,. \tag{1.23}$$

From this various relations between thermodynamic derivatives can be deduced. For example, it follows immediately from Eq. (1.23) that

$$T = \left(\frac{\partial U}{\partial S} \right)_V \text{ and } -p = \left(\frac{\partial U}{\partial V} \right)_S \,; \tag{1.24}$$

but a property of partial differentiation is that $\partial^2 f / \partial x \partial y = \partial^2 f / \partial y \partial x$, so we find that

$$\left(\frac{\partial T}{\partial V} \right)_S = -\left(\frac{\partial p}{\partial S} \right)_V \,. \tag{1.25}$$

Another useful manipulation that is a general property of partial derivatives is that

$$\frac{\partial f / \partial y)_x}{(\partial f / \partial x)_y} \equiv -\left(\frac{\partial x}{\partial y} \right)_f \,. \tag{1.26}$$

This follows by rearranging $\mathrm{d}f = (\partial f / \partial x)_y \, \mathrm{d}x + (\partial f / \partial y)_x \, \mathrm{d}y$ to make $\mathrm{d}x$ the subject of the formula, and identifying the resulting coefficient of $\mathrm{d}y$ as the derivative $(\partial x / \partial y)_f$. Various thermodynamic relations that are useful

in stellar physics and astrophysical fluid dynamics can be found in e.g.
Kippenhahn & Weigert (1990).

We define the *adiabatic exponents* γ_1, γ_2, γ_3 by

$$\gamma_1 = \left(\frac{\partial \ln p}{\partial \ln \rho}\right)_S, \quad \frac{\gamma_2 - 1}{\gamma_2} = \left(\frac{\partial \ln T}{\partial \ln p}\right)_S, \quad \gamma_3 - 1 = \left(\frac{\partial \ln T}{\partial \ln \rho}\right)_S.$$

$$(1.27)$$

Note that all these partial derivatives are at constant specific entropy: 'adiabatic' here means without exchange of heat, so $\delta Q = 0 = dS$, i.e. S is constant. The quantity $(\gamma_2 - 1)/\gamma_2 \equiv (\partial \ln T/\partial \ln p)_S$ is often referred to as ∇_{ad}.

We define c_p, the specific heat at constant pressure, to be the amount of heat required to make a unit increase in temperature, without the pressure changing: thus, from Eq. (1.22), $c_p = T(\partial S/\partial T)_p$. Similarly we define c_V, the specific heat at constant volume, to be the amount of heat required to make a unit increase in temperature at constant V. The following three useful results relate the amount of heat added to the changes in pairs of thermodynamic variables:

$$\delta Q = \frac{1}{\rho(\gamma_3 - 1)}\left(dp - \frac{\gamma_1 p}{\rho}d\rho\right)$$

$$\delta Q = c_p\left(dT - \frac{\gamma_2 - 1}{\gamma_2}\frac{T}{p}dp\right) \qquad (1.28)$$

$$\delta Q = c_V\left(dT - (\gamma_3 - 1)\frac{T}{\rho}d\rho\right).$$

For example, the first equation can be derived by noting

$$\delta Q \equiv TdS = T\left(\frac{\partial S}{\partial p}\right)_V dp + T\left(\frac{\partial S}{\partial V}\right)_p dV$$

$$= T\left(\frac{\partial p}{\partial S}\right)_V^{-1}\left[dp + T\frac{(\partial S/\partial V)_p}{(\partial S/\partial p)_V}dV\right]$$

and using Eq. (1.25) on the factor outside the square brackets and Eq. (1.26) to manipulate the last term, together with the definitions of γ_1 and γ_3. Note that $\ln V = -\ln \rho$. The other two above expressions for δQ are derived similarly.

1.7   Perfect Gases

A perfect gas is one for which

$$pV = RT \qquad (1.29)$$

(R being some constant), and

$$U = U(T). \qquad (1.30)$$

It follows from Eq. (1.29) that

$$\frac{dp}{p} + \frac{dV}{V} = \frac{dT}{T}. \qquad (1.31)$$

Now for an adiabatic change of such a gas,

$$0 = dS = \frac{1}{T}\left(dU + pdV\right) = \frac{dU}{dT}\frac{dT}{T} + R\frac{dV}{V}. \qquad (1.32)$$

From Eq. (1.32) and the definition of γ_3 it follows that

$$\gamma_3 = 1 + \frac{R}{dU/dT}. \qquad (1.33)$$

Eliminating dT between Eqs. (1.31) and (1.32) gives that γ_1 is given by the same expression (1.33), and likewise for γ_2 (eliminating dV). Thus for a perfect gas, the three adiabatic exponents are equal.

Henceforward in this book, since for a perfect gas all three adiabatic exponents are equal, we shall use γ to denote all of them when no confusion can arise.

In fact, for a monatomic gas (in which the molecules are simply point masses) one can show that γ is equal to 5/3, as follows. For a monatomic gas, the internal energy is just the translational kinetic energy of all the molecules. Assuming the gas to be isotropic (all directions equivalent) and all the molecules identical, the total internal energy of the gas in volume V is

$$U = \frac{1}{2}Nm\left(\overline{v_x^2} + \overline{v_y^2} + \overline{v_z^2}\right) = \frac{3}{2}Nm\overline{v_x^2}, \qquad (1.34)$$

where m is the mass of a molecule, N is the number of molecules in V, and (for example) $\overline{v_x^2}$ is the mean squared velocity in the x direction. Suppose the volume V is enclosed by a rigid rectangular box of length l in the x-direction (and of cross-sectional area $A = V/l$). The force on the end of the box is pA. Consider a single molecule. It has some x-velocity v_x. In time $\Delta t \equiv 2l/v_x$ it bounces off that end of the box once. In bouncing, its x-momentum changes by an amount $2mv_x$ (assuming an elastic collision). Thus, since force (= pressure × area) is equal to the rate of change of momentum, summing over all molecules gives

$$pA = \sum \frac{2mv_x}{\Delta t} = \sum \frac{m}{l} v_x^2 = \frac{Nm}{l} \overline{v_x^2} . \qquad (1.35)$$

Hence $pV = Nm\overline{v_x^2}$ and so, from Eqs. (1.29), (1.33) and (1.34),

$$U = \frac{3}{2} RT \quad \text{and} \quad \gamma = \frac{5}{3} . \qquad (1.36)$$

If a gas is undergoing ionization, dU/dT is greater than it would otherwise be, because energy goes into ionizing the gas; so from e.g. Eq. (1.33) the adiabatic exponents are reduced in value.

1.8   The Virial Theorem

The velocity u is the rate of change of position following the fluid:

$$u = \frac{Dr}{Dt} . \qquad (1.37)$$

Hence Eq. (1.7), with f replaced by gravitational acceleration and using Eq. (1.12), can be rewritten

$$\rho \frac{D^2 r}{Dt^2} = -\nabla p - \rho \nabla \psi . \qquad (1.38)$$

Taking the dot product with r and integrating over the whole volume of the fluid gives

$$\int_V r \cdot \frac{D^2 r}{Dt^2} \rho \, dV = -\int_V r \cdot \nabla p \, dV - \int_v r \cdot \nabla \psi \, \rho \, dV . \qquad (1.39)$$

The left-hand side of Eq. (1.39) can be rewritten as

$$\frac{d}{dt} \int_V r \cdot \frac{Dr}{Dt} \rho \, dV - \int_V \left(\frac{Dr}{Dt} \right)^2 \rho \, dV = \frac{1}{2} \frac{d^2}{dt^2} \int_V |r|^2 \rho \, dV - 2\mathcal{T} , \quad (1.40)$$

where $\mathcal{T} \equiv \frac{1}{2} \int_V \rho u^2 \, dV$ is the total kinetic energy of the fluid. (Note that here and in similar expressions we write u^2 when what is meant is $|u|^2$, i.e. $u \cdot u$ — the quantity is a scalar.)

Using the divergence theorem and the identity $\nabla \cdot \boldsymbol{r} \equiv \partial x_i / \partial x_i = 3$, the pressure term in Eq. (1.39) can be rewritten as

$$-\int_V \boldsymbol{r} \cdot \nabla p \, dV = -\int_S p \boldsymbol{r} \cdot \boldsymbol{n} dS + 3 \int_V p \, dV . \tag{1.41}$$

We suppose that the pressure vanishes at the boundary of the fluid volume (this can be a good approximation for a star, for example) so that the surface term is zero.

Finally,

$$\begin{aligned}
-\int_V \boldsymbol{r} \cdot \nabla \psi \, \rho dV &= G \int_V \int_{V'} \boldsymbol{r} \cdot \nabla \Big(\frac{\rho(\boldsymbol{r}')}{|\boldsymbol{r} - \boldsymbol{r}'|} \Big) \rho(\boldsymbol{r}) \, dV \, dV' \\
&= -G \int_V \int_{V'} \frac{\boldsymbol{r} \cdot (\boldsymbol{r} - \boldsymbol{r}')}{|\boldsymbol{r} - \boldsymbol{r}'|^3} \rho(\boldsymbol{r}) dV \, \rho(\boldsymbol{r}') dV' \\
&= -\frac{1}{2} G \int_V \int_{V'} \Big(\frac{\boldsymbol{r} \cdot (\boldsymbol{r} - \boldsymbol{r}')}{|\boldsymbol{r} - \boldsymbol{r}'|^3} + \frac{\boldsymbol{r}' \cdot (\boldsymbol{r}' - \boldsymbol{r})}{|\boldsymbol{r} - \boldsymbol{r}'|^3} \Big) \rho(\boldsymbol{r}) dV \, \rho(\boldsymbol{r}') dV' \\
&= -\frac{1}{2} G \int_V \int_{V'} \frac{1}{|\boldsymbol{r} - \boldsymbol{r}'|} \rho(\boldsymbol{r}) dV \, \rho(\boldsymbol{r}') dV' \\
&= \Psi
\end{aligned} \tag{1.42}$$

(the later steps exploit the symmetry between $\boldsymbol{r}$ and $\boldsymbol{r}'$) where

$$\Psi \equiv \frac{1}{2} \int_V \psi \, \rho(\boldsymbol{r}) dV \tag{1.43}$$

is the total gravitational energy. Putting all this together yields

$$\frac{1}{2} \frac{d^2 \mathcal{I}}{dt^2} = 2\mathcal{T} + 3 \int_V p \, dV + \Psi , \tag{1.44}$$

where $\mathcal{I} \equiv \int_V \rho r^2 \, dV$. Equation (1.44) is the scalar form of the *virial theorem*.

One can also derive a tensor virial theorem, by taking the ith component of Eq. (1.38) and multiplying by the jth component of $\boldsymbol{r}$:

$$\rho x_j \frac{D^2 x_i}{Dt^2} = -x_j \frac{\partial p}{\partial x_i} - \rho x_j \frac{\partial \psi}{\partial x_i} . \tag{1.45}$$

It can then be shown that

$$\frac{1}{2} \frac{d^2 I_{ij}}{dt^2} = 2T_{ij} + \delta_{ij} \int_V p \, dV + \Psi_{ij} , \tag{1.46}$$

where

$$I_{ij} = \int_V \rho x_i x_j dV ,$$

$$T_{ij} = \frac{1}{2} \int_V \rho u_i u_j dV ,$$

(1.47)

$$\Psi_{ij} = -\frac{1}{2} G \int_V \int_{V'} \frac{(x_i - x_i')(x_j - x_j')}{|r - r'|^3} \rho(r) dV \, \rho(r') dV' .$$

Note that if Eq. (1.46) is contracted over i and j (i.e. multiplied by δ_{ij} and summed over i and j) then the scalar virial theorem (1.44) is recovered.

Derivations of the virial theorem in different forms can be found in Chandrasekhar (1969) and Tassoul (1978).

1.9   Vorticity

An important derived quantity for a fluid flow is the vorticity

$$\boldsymbol{\omega} = \nabla \times \boldsymbol{u} .$$

(1.48)

For a fluid rotating rigidly with angular velocity $\boldsymbol{\Omega}$, for example, $\boldsymbol{u} = \boldsymbol{\Omega} \times \boldsymbol{r}$ and

$$\boldsymbol{\omega} = 2\boldsymbol{\Omega} ,$$

(1.49)

using a standard vector identity (cf. Appendix A). Generally, in a fluid flow $\boldsymbol{u}$ the vorticity at any location is equal to twice the local rotation rate of a fluid line element at that location, as is proved below. This does not mean however that streamlines have to be curved for the fluid to possess non-zero vorticity. For example, consider the shear flow $\boldsymbol{u} = Cy\boldsymbol{e}_x$ where C is a non-zero constant: this is a unidirectional shear flow in the x-direction with magnitude proportional to y. Here as elsewhere we use $\boldsymbol{e}_x$, $\boldsymbol{e}_y$, $\boldsymbol{e}_z$ to denote unit vectors in the x-, y- and z-directions. It is a straightforward exercise to show that the vorticity of such a flow is $\boldsymbol{\omega} = -C\boldsymbol{e}_z$, which is non-zero although the streamlines (lines everywhere parallel to the flow) are straight lines.

To see the relationship between vorticity and local rotation of the fluid, we shall now analyse the relative motion of a fluid in the vicinity of a point. Let A be a point moving with the fluid, which at an initial time t is at position r; and let B be another point which at time t is at nearby position $r + \delta r$ (Fig. 1.2). At time t therefore the position of B relative to A is δr.

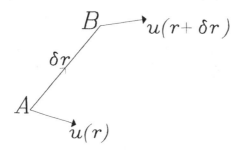

Fig. 1.2 The motion of neighbouring points A (at position r) and B (at postion $r + \delta r$), which leads to the evolution of the material line element δr with time.

The fluid velocity at A is $u(r)$ and at B it is $u(r + \delta r)$; therefore after a short time δt the separation of B from A has changed to

$$\{r + \delta r + \delta t\, u(r + \delta r)\} \;-\; \{r + \delta t\, u(r)\} \;\equiv\; \delta r + \delta t\, \{u(r + \delta r) - u(r)\}$$

correct to $O(\delta t)$. The last term can be simplified by expanding $u(r + \delta r)$ in a Taylor series about r and keeping only terms up to δr. Treating δr now as a function of time, an equation for the rate of change of δr with time can be can be obtained from here by dividing the difference between the new separation and the old one by δt and taking the limit as $\delta t \to 0$:

$$\frac{D\delta r}{Dt} \;=\; \delta r \cdot \nabla u \tag{1.50}$$

where, since we are following the separation between material points, we write the derivative as a material derivative. This equation therefore describes the evolution of a material element δr. The right-hand side can be expressed in index notation as $(\delta r)_j \partial u_i / \partial x_j$. The tensor ∇u, like any other second-rank tensor, can be split into a symmetric part and an anti-symmetric part:

$$\frac{\partial u_i}{\partial x_j} \;\equiv\; \frac{1}{2}\left(\frac{\partial u_i}{\partial x_j} + \frac{\partial u_j}{\partial x_i}\right) + \frac{1}{2}\left(\frac{\partial u_i}{\partial x_j} - \frac{\partial u_j}{\partial x_i}\right). \tag{1.51}$$

The second term on the right of Eq. (1.51) is anti-symmetric: using the definition (1.48) of vorticity, it can be written as $-\frac{1}{2}\epsilon_{ijk}\omega_k$ (cf. Appendix A). Hence, substituting into Eq. (1.50) it makes a contribution to the right-hand side which is equal to $-\frac{1}{2}\delta r \times \omega \equiv \frac{1}{2}\omega \times \delta r$. Thus, comparing this with the velocity due to a solid-body rotation, we deduce that the antisymmetric part of ∇u contibutes to the motion of point B relative to point A a motion which is a rotation with angular velocity $\frac{1}{2}\omega$. Equivalently, the vorticity is given by Eq. (1.49) where $\mathbf{\Omega}$ is interpreted as the local rotation rate.

The symmetric part of $\partial u_i/\partial x_j$, i.e. the first term on the right of Eq. (1.51), is called the rate of strain tensor e_{ij}:

$$e_{ij} = \frac{1}{2}\left(\frac{\partial u_i}{\partial x_j} + \frac{\partial u_j}{\partial x_i}\right). \tag{1.52}$$

Its trace e_{kk} is equal to $\nabla \cdot u$ and the isostropic part of e_{ij}, namely $\frac{1}{3}e_{kk}\delta_{ij}$, represents an expansion or compression of the fluid in the region of point A. The remainder of e_{ij}, namely $e_{ij} - \frac{1}{3}e_{kk}\delta_{ij}$, has zero trace and represents a local shear of the fluid.

An evolution equation for vorticity can be derived by taking the curl of the momentum equation. Initially we shall consider the inviscid case. First we rewrite the $u \cdot \nabla u$ term in Eq. (1.7) using a vector identity to obtain

$$\frac{\partial u}{\partial t} = u \times \omega - \nabla\left(\frac{1}{2}u^2\right) - \frac{1}{\rho}\nabla p + f. \tag{1.53}$$

Taking the curl of this equation gives

$$\frac{\partial \omega}{\partial t} = \nabla \times (u \times \omega) + \frac{1}{\rho^2}\nabla\rho \times \nabla p + \nabla \times f \tag{1.54}$$

since the curl of a gradient vector is zero. The $\nabla \times (u \times \omega)$ can be expanded using identity (A.4) from Appendix A. Noting that $\nabla \cdot \omega$ vanishes by its definition (1.48), Eq. (1.54) becomes

$$\frac{\partial \omega}{\partial t} + u \cdot \nabla\omega = \omega \cdot \nabla u - (\nabla \cdot u)\omega + \frac{1}{\rho^2}\nabla\rho \times \nabla p + \nabla \times f. \tag{1.55}$$

Finally, using the continuity equation (1.3) to eliminate $\nabla \cdot u$, we obtain

$$\frac{D}{Dt}\left(\frac{\omega}{\rho}\right) \equiv \frac{\partial}{\partial t}\left(\frac{\omega}{\rho}\right) + u \cdot \nabla\left(\frac{\omega}{\rho}\right) = \left(\frac{\omega}{\rho}\right) \cdot \nabla u + \frac{1}{\rho^3}\nabla\rho \times \nabla p + \frac{1}{\rho}\nabla \times f. \tag{1.56}$$

Equation (1.56) is called the *vorticity equation*. It describes how vorticity evolves in a fluid.

A fluid for which $\nabla\rho \times \nabla p = 0$ everywhere is called *barotropic*: since the vector gradients of density and pressure are everywhere parallel, the surfaces of constant density and of constant pressure coincide, and it is possible to write either variable as a function solely of the other variable, e.g. $\rho = \rho(p)$. Conversely, if density and pressure are just functions one of the other, then the fluid is barotropic. If the fluid is barotropic and any body force $\boldsymbol{f}$ that is present is conservative (i.e. $\nabla \times \boldsymbol{f} = 0$, or equivalently $\boldsymbol{f}$ can be written as the gradient of a scalar potential), so in particular there are no viscous forces, then the last two terms in Eq. (1.56) vanish and the vorticity equation (1.56) is of the same form as the equation (1.50) for the evolution of a material line element. We deduce that in this case vortex lines move with the fluid. (A vortex line is a line everywhere parallel to the vorticity.) We can define a *vortex tube* to be, loosely speaking, a bundle of vortex lines or more precisely a tube whose surface is nowhere crossed by vortex lines and whose surface is itself composed of vortex lines. In the present case, then, the walls of a vortex tube form a material surface moving with the fluid. We define the strength of a vortex tube to be $\int_S \boldsymbol{\omega} \cdot \boldsymbol{n} \, dS$, where S is any cross-section area cut across the tube and $\boldsymbol{n}$ is a vector normal to that area. Since no vortex lines cross the walls of a vortex tube, and vorticity is divergence-free, it follows from the divergence theorem that the strength of a vortex tube is a well-defined quantity, i.e. it is independent of which cross-section area we choose on which to evaluate it.

Generally in a fluid we can define the *circulation* about a closed curve $\mathcal{C}$ contained within the fluid to be

$$\Gamma = \oint_{\mathcal{C}} \boldsymbol{u} \cdot d\boldsymbol{r} \, . \tag{1.57}$$

If we choose $\mathcal{C}$ to be a material curve, moving with the fluid, then

$$\frac{d\Gamma}{dt} = \oint_{\mathcal{C}} \frac{D\boldsymbol{u}}{Dt} \cdot d\boldsymbol{r} + \oint_{\mathcal{C}} \boldsymbol{u} \cdot \frac{D}{Dt} d\boldsymbol{r} \, . \tag{1.58}$$

The last term can be rewritten as the integral around $\mathcal{C}$ of $\nabla(\frac{1}{2}\boldsymbol{u}^2)$ using Eq. (1.50), and for an inviscid fluid we can replace $D\boldsymbol{u}/Dt$ in the first term on the right of Eq. (1.58) using the momentum equation (1.7). We consider only 3-D fluid domains for which curve $\mathcal{C}$ can be spanned by a surface $\mathcal{S}$ wholly contained in the fluid domain; so Stokes's theorem can be applied. Hence Eq. (1.58) becomes

$$\frac{d\Gamma}{dt} = \int_{\mathcal{S}} \left(\frac{1}{\rho^2} \nabla\rho \times \nabla p + \nabla \times \boldsymbol{f} \right) \cdot \boldsymbol{n} \, dS \, . \tag{1.59}$$

We can immediately deduce from this that for a barotropic fluid with only conservative body forces f, the circulation around a material curve is invariant with time. This is a statement of *Kelvin's circulation theorem*.

Moreover, the strength of a vortex tube can be expressed, using Stokes's theorem, as a flow around a material curve embedded in the walls of the tube and encircling the tube's axis. Hence the strength of a vortex tube in such a flow is invariant also: if the fluid motion is such as to cause the vortex tube to become narrower (known as vortex stretching), the magnitude of ω must increase so that $\int \omega \cdot n \, dS$ over the cross-section of the tube is constant.

A fluid for which $\nabla \rho \times \nabla p \neq 0$ is called *baroclinic*. This means that surfaces of constant density are inclined to the surfaces of constant pressure.

Chapter 2

Simple Models of Astrophysical Fluids and Their Motions

In the first chapter we established the momentum equation (1.7), the continuity equation (1.4), Poisson's equation (1.15) and the energy equation (1.19). Assuming that the only body forces are due to self-gravity, so that $\boldsymbol{f} = -\nabla\psi$ in Eq. (1.7), these equations are:

$$\rho\frac{D\boldsymbol{u}}{Dt} = -\nabla p - \rho\nabla\psi\,, \tag{2.1}$$

$$\frac{D\rho}{Dt} + \rho\operatorname{div}\boldsymbol{u} = 0\,, \tag{2.2}$$

$$\nabla^2\psi = 4\pi G\rho\,, \tag{2.3}$$

$$\frac{DU}{Dt} - \frac{p}{\rho^2}\frac{D\rho}{Dt} = \epsilon - \frac{1}{\rho}\nabla\cdot\boldsymbol{F}\,. \tag{2.4}$$

Note that these contain seven dependent variables, namely ρ, the three components of $\boldsymbol{u}$, p, ψ and U. The three components of Eq. (2.1), together with Eqs. (2.2)–(2.4), provide six equations, and a seventh is the equation of state (e.g. that for a perfect gas) which provides a relation between any three thermodynamic state variables, so that (for example) the internal energy U and temperature T can be written in terms of p and ρ. (It is assumed that ϵ and $\boldsymbol{F}$ are known functions of the other variables.) Thus one might hope in principle to solve these equations, given suitable boundary conditions. In practice this set of equations is intractable to exact solution, and one must resort to numerical solutions. Even these can be extremely problematic so that, for example, understanding turbulent flows is still a very challenging research area. Moreover, an analytic solution to a somewhat idealized problem may teach one much more than a numerical

solution. One useful idealization is where we assume that the fluid velocity
and all time derivatives are zero. These are called equilibrium solutions
and describe a steady state. Although a true steady state may be rare in
reality, the time-scale over which an astrophysical system evolves may be
very long, so that at any particular time the state of many astrophysical
fluid bodies such as stars may be well represented by an equilibrium model.
Even when the dynamical behaviour of the body is important, it can often
be described in terms of small departures from an equilibrium state. Hence
in this chapter we start by looking at some equilibrium models and then
derive equations describing small perturbations about an equilibrium state.

2.1 Hydrostatic Equilibrium for a Self-gravitating Body

If we suppose that $u = 0$ everywhere, and that all quantities are independent of time, then Eq. (2.1) becomes

$$\nabla p \ = \ \rho g \ = \ - \rho \nabla \psi \ ; \tag{2.5}$$

the continuity equation becomes trivial; and Eq. (2.3) is unchanged. A
fluid satisfying Eq. (2.5) is said to be in hydrostatic equilibrium. If it is
self-gravitating (so that ψ is determined by the density distribution within
the fluid), then Eq. (2.3) must also be satisfied.

Putting $u = 0$ and $\partial/\partial t = 0$ in Eq. (2.4), we obtain that the heat sources
given by ϵ must be exactly balanced by the heat flux term $\rho^{-1}\nabla \cdot F$. If
this holds, then the fluid is also said to be in thermal equilibrium. Since
we have not yet considered what the heat sources might be, nor the details
of the heat flux, we shall neglect considerations of thermal equilibrium at
this point.

2.1.1 *Spherically symmetric case*

In spherical polar coordinates (r, θ, ϕ) (see Appendix B),

$$\nabla^2\psi \ = \ \frac{1}{r^2}\frac{\partial}{\partial r}\left(r^2\frac{\partial\psi}{\partial r}\right) + \frac{1}{r^2\sin\theta}\frac{\partial}{\partial\theta}\left(\sin\theta\frac{\partial\psi}{\partial\theta}\right) + \frac{1}{r^2\sin^2\theta}\frac{\partial^2\psi}{\partial\phi^2} \ . \tag{2.6}$$

Let us seek a solution where everything is independent of θ and ϕ (and
hence dependent only on the radial variable r). Then Eq. (2.3) becomes

$$\frac{1}{r^2}\frac{\mathrm{d}}{\mathrm{d}r}\left(r^2\frac{\mathrm{d}\psi}{\mathrm{d}r}\right) \ = \ 4\pi G\rho(r) \ . \tag{2.7}$$

Integrating once gives

$$r^2 \frac{\mathrm{d}\psi}{\mathrm{d}r} = Gm(r) \tag{2.8}$$

where

$$m(r) = \int_0^r 4\pi \tilde{r}^2 \rho(\tilde{r}) \, \mathrm{d}\tilde{r} . \tag{2.9}$$

(Note that $m(\mathrm{r})$ is the mass inside a sphere of radius r, centred on the origin.) Now $\nabla \psi = (\mathrm{d}\psi/\mathrm{d}r)e_r$ if ψ is only a function of r, e_r being a unit vector in the radial direction. Hence Eq. (2.8) implies

$$g \equiv -\nabla \psi = -\frac{Gm}{r^2} e_r . \tag{2.10}$$

Equation (2.10) states that in the spherically symmetric case, the gravitational acceleration at position r is due only to the mass interior to r and independent of the density distrubtion outside r: this is known as Newton's sphere theorem. Also, by Eq. (2.5),

$$\nabla p = -\frac{Gm\rho}{r^2} e_r . \tag{2.11}$$

The vector ∇p points towards the origin, so the pressure decreases as r increases.

One can only make further progress by assuming some relation between pressure and density. Suppose then that the fluid is a perfect gas, so

$$p = \frac{\mathcal{R}\rho T}{\mu} \equiv a^2 \rho ; \tag{2.12}$$

a is known as the isothermal sound speed. Suppose further that the temperature T and mean molecular weight μ are both constants throughout the fluid, so a is also a constant. Then Eq. (2.11) becomes

$$a^2 \frac{\mathrm{d}\rho}{\mathrm{d}r} = -\frac{Gm\rho}{r^2} ,$$

which implies that

$$\frac{\mathrm{d}}{\mathrm{d}r} \left(\frac{r^2 a^2}{\rho} \frac{\mathrm{d}\rho}{\mathrm{d}r} \right) = -4\pi G r^2 \rho . \tag{2.13}$$

Seeking a solution of the form $\rho = Ar^n$, where A and n are constants, gives

$$\rho = \frac{a^2}{2\pi G r^2} , \quad p = \frac{a^4}{2\pi G r^2} . \tag{2.14}$$

This is the singular self-gravitating isothermal sphere solution. It is not physically realistic at $r = 0$, where p and ρ are singular, but nonetheless it is a useful analytical model solution. Of course, in a real nondegenerate star, for example, the interior is not isothermal: the temperature increases with depth, which in turn means that the pressure increases and the star is prevented from collapsing in upon itself without recourse to infinite pressure and density at the centre.

2.1.2 *Plane-parallel layer under constant gravity*

In modelling the atmosphere and outer layers of a star, the spherical geometry can often be ignored, so that such a region can be approximated as a plane-parallel layer. Moreover, in the rarified outer layers of a star the gravitational acceleration $\boldsymbol{g}$ may be approximated as a constant vector. Thus, in Cartesian coordinates (x, y, z) we have a region in which everything is a function of z alone and $\boldsymbol{g} = -g\boldsymbol{e}_z$, where g is constant. Note that we take z to be *height*, so $\boldsymbol{e}_z$ points upwards. Hence Eq. (2.5) becomes

$$\frac{\mathrm{d}p}{\mathrm{d}z} = -g\,\rho(z) . \qquad (2.15)$$

Since self-gravity is being ignored, Eq. (2.3) is not used.

In the *isothermal* case ($p/\rho = a^2$ constant), Eq. (2.15) can be integrated to give

$$\rho = \rho_0 \exp(-gz/a^2) \qquad (2.16)$$

where the constant ρ_0 is the density at $z = 0$.

The density scale height H is defined by

$$H = \left| \frac{1}{\rho} \frac{\mathrm{d}\rho}{\mathrm{d}z} \right|^{-1} . \qquad (2.17)$$

Hence, in this case, $H = a^2/g$ and is constant. Thus $\rho = \rho_0 \exp(-z/H)$.

A useful family of solutions is that of *plane-parallel polytropes*, where

$$p = K\rho^{1+1/n} \qquad (2.18)$$

(with both K and n constant); n is called the polytropic index. For example, in the adiabatically stratified convection zone of the Sun the pressure–density relation is well described by $p = K\rho^\gamma$ where $\gamma = 5/3$ (except in regions of partial ionization) and hence, comparing with Eq. (2.18) by a polytrope of index $n = 1.5$. Note that the isothermal layer is obtained in the limit $n \to \infty$.

Substituting Eq. (2.18) into Eq. (2.15) gives

$$K \frac{(n+1)}{n} \rho^{\frac{1}{n} - 1} \frac{d\rho}{dz} = -g \qquad (2.19)$$

and this integrates to give

$$\rho^{1/n} = \frac{-gz}{(n+1)K} + \text{constant} . \qquad (2.20)$$

If the density vanishes at $z = 0$ (which could be a reasonable approximation if $z = 0$ were the surface of a star) then the constant of integration in Eq. (2.20) is zero. Hence for a plane-parallel polytrope of finite index n, $\rho \propto (-z)^n$ and $p \propto) - z)^{n+1}$; also $T \propto p/\rho \propto -z$, so the temperature increases linearly with depth.

2.2 Equations of Stellar Structure

Although it is an aside, it may be instructive to point out here the relationship between the fluid equations that we have derived thus far and the equations of stellar structure describing a static, spherically symmetric star. These are commonly used in studies of stellar structure and evolution.

The equation of hydrostatic equilibrium is just the momentum equation in the static case: with spherical symmetry, so that quantities are only functions of radial variable r, this is given by Eq. (2.11):

$$\frac{dp}{dr} = -\frac{Gm\rho}{r^2} . \qquad (2.21)$$

A differential equation for variation of mass $m(r)$ contained within a sphere of radius r is just the derivative of Eq. (2.9):

$$\frac{dm}{dr} = 4\pi r^2 \rho . \qquad (2.22)$$

In a spherically symmetric star the heat flux $\boldsymbol{F}$ is purely radial: $\boldsymbol{F} = F(r)\boldsymbol{e}_r$. The flux $F(r)$ is related to the total luminosity through a sphere of radius r by $L(r) = 4\pi r^2 F(r)$. An equation describing the radial variation of luminosity follows from Eq. (2.4). Setting the derivatives on the left of that equation to zero, and using the expression for divergence in spherical polar coordinates (Appendix B), it follows that

$$\frac{dL}{dr} = 4\pi r^2 \rho \epsilon \,. \tag{2.23}$$

A fourth and final differential equation describes how heat is transported in the star. In the bulk of a star like the Sun this is by radiation. To combine radiative transfer with fluid dynamics in general is a substantial topic in itself, and is excellently expounded by Mihalas & Mihalas (1984). Here we only consider a static case and moreover, in the interior of a star the radiative transport is well described by a diffusion equation. The prototypical diffusive transport equation for the flux $\boldsymbol{j}_q$ of some quantity q with density ρ_q per unit volume is

$$\boldsymbol{j}_q = -D\nabla\rho_q \,, \tag{2.24}$$

where D is the coefficient of diffusion. Typically D can be related to the mean free path l amd mean speed v of particles carrying the flux, by

$$D = \frac{1}{3}v\,l \,. \tag{2.25}$$

In the present case of radiative diffusion of heat, the particle velocity is the speed of light (which in this subsection only we denote by c), and astrophysicists describe the mean free path in terms of a mean opacity to radiation (κ), thus $l = (\kappa\rho)^{-1}$. The density of thermal energy in the radiation is $U = aT^4$, where a is the radiation constant. Thus finally we obtain the fourth differential equation after some rearrangement to make dT/dr the subject of the formula as

$$\frac{dT}{dr} = \frac{-3\,\kappa\,\rho\,L}{16\pi acr^2 T^3} \tag{2.26}$$

in regions of the star where heat is transported solely by radiation. In convectively unstable regions the heat transport is by convection (or some combination of convection and radiation). If convection is very efficient, this typically leads to a stratification that is very close to marginal stability (see Section 4.1.1) in which the temperature gradient is instead given to a very good approximation by

$$\frac{dT}{dr} = \left(1 - \frac{1}{\gamma}\right)\frac{T}{p} \,. \tag{2.27}$$

More details of the derivation and use of these equations, and of stellar structure and evolution in general, may be found in e.g. the book by Kippenhahn & Weigert (1990). It should be evident from the above discussion of the origins of the standard equations of stellar structure that, if the star is not static or not spherically symmetric, it is appropriate to return to the full fluid dynamical equations to obtain equations appropriate for modelling the star.

2.3 Small Perturbations about Equilibrium

In many interesting instances, the motion of a fluid body may be considered to be a small disturbance about an equilibrium state. Suppose that in equilibrium the pressure, density and gravitational potential are given by $p = p_0$, $\rho = \rho_0$, $\psi = \psi_0$ (all possibly functions of position, but independent of time) and $u = 0$. Using Eqs. (2.5) and (2.3), the equilibrium quantities satisfy

$$\nabla p_0 = -\rho_0 \nabla \psi_0 , \qquad \nabla^2 \psi_0 = 4\pi G \rho_0 . \tag{2.28}$$

Suppose now that the system undergoes small motions about the equilibrium state, so

$$p = p_0 + p' , \quad \rho = \rho_0 + \rho' , \quad \psi = \psi_0 + \psi' , \tag{2.29}$$

so for example $p'(\boldsymbol{r}, t) \equiv p(\boldsymbol{r}, t) - p_0(\boldsymbol{r})$ is the difference between the actual pressure at time t and position $\boldsymbol{r}$ and its equilibrium value there. Substituting these expressions into Eqs. (2.1) – (2.3) yields

$$(\rho_0 + \rho')\Big(\frac{\partial \boldsymbol{u}}{\partial t} + \boldsymbol{u} \cdot \nabla \boldsymbol{u}\Big) = -\nabla(p_0 + p') - (\rho_0 + \rho')\nabla(\psi_0 + \psi') ,$$

$$\frac{\partial}{\partial t}(\rho_0 + \rho') = -\nabla \cdot \big((\rho_0 + \rho')\boldsymbol{u}\big) , \tag{2.30}$$

$$\nabla^2(\psi_0 + \psi') = 4\pi G(\rho_0 + \rho') .$$

We suppose that the perturbations (the primed quantities and the velocity) are small; hence we neglect the products of two or more small quantities, since these will be even smaller. This is known as linearization, because we only retain equilibrium terms and terms that are linear in small quantities.

This simplifies the above equations to

$$\rho_0 \frac{\partial \boldsymbol{u}}{\partial t} = -\nabla(p_0 + p') - (\rho_0 + \rho')\nabla\psi_0 - \rho_0\nabla\psi',$$

$$\frac{\partial \rho'}{\partial t} = -\nabla \cdot (\rho_0 \boldsymbol{u}), \tag{2.31}$$

$$\nabla^2(\psi_0 + \psi') = 4\pi G(\rho_0 + \rho').$$

Subtracting equilibrium Eqs. (2.28) leaves a set of equations all the terms of which are linear in small quantities:

$$\rho_0 \frac{\partial \boldsymbol{u}}{\partial t} = -\nabla p' - \rho'\nabla\psi_0 - \rho_0\nabla\psi', \tag{2.32}$$

$$\frac{\partial \rho'}{\partial t} = -\nabla \cdot (\rho_0 \boldsymbol{u}), \tag{2.33}$$

$$\nabla^2\psi' = 4\pi G\rho'. \tag{2.34}$$

Equations (2.32)–(2.34) give five equations (counting the vector equation as three) for six unknowns ($\boldsymbol{u}$, p', ρ', ψ'). We need another equation to close the system: that equation comes from energy considerations. In generality, we should perturb the energy equation (2.4) in the same manner as Eqs. (2.1)–(2.3). But there are two limiting cases, isothermal perturbations and adiabatic perturbations, which are sufficiently common to be very useful and are simpler than using the full perturbed equation (2.4) because they don't involve a detailed description of how ϵ and $\boldsymbol{F}$ are perturbed.

2.3.1 *Isothermal fluctuations*

Let the typical time scale and length scale on which the perturbations vary be τ and λ, respectively. Suppose that the timescale on which heat can be exchanged over a distance λ is much shorter than τ. Since heat tends to flow from hotter regions to cooler ones, efficient heat exchange will eliminate any temperature fluctuations. Assuming a perfect gas, perturbing equation (2.12) gives

$$\frac{\mathrm{d}p}{p} = \frac{\mathrm{d}\rho}{\rho} + \frac{\mathrm{d}T}{T}. \tag{2.35}$$

For isothermal fluctuations, $\mathrm{d}T = 0$. Hence $\mathrm{d}p/p = \mathrm{d}\rho/\rho$. In terms of material derivatives,

$$\frac{Dp}{Dt} = \frac{p}{\rho}\frac{D\rho}{Dt}. \tag{2.36}$$

The linearized form of this equation is

$$\frac{\partial p'}{\partial t} + \boldsymbol{u} \cdot \nabla p_0 = \frac{p_0}{\rho_0} \left(\frac{\partial \rho'}{\partial t} + \boldsymbol{u} \cdot \nabla \rho_0 \right) . \qquad (2.37)$$

2.3.2 Adiabatic fluctuations

The converse situation is where the timescale for heat exchange between neighbouring material is much longer than the timescale of the perturbations. Then we can say that over a timescale τ the heat gained or lost by a fluid element is zero: $\delta Q = 0$. By Eq. (1.28) this implies that

$$\frac{dp}{p} = \gamma \frac{d\rho}{\rho} , \qquad (2.38)$$

or in terms of material derivatives

$$\frac{Dp}{Dt} = \frac{\gamma p}{\rho} \frac{D\rho}{Dt} . \qquad (2.39)$$

The linearized form of this equation is

$$\frac{\partial p'}{\partial t} + \boldsymbol{u} \cdot \nabla p_0 = \frac{\gamma p_0}{\rho_0} \left(\frac{\partial \rho'}{\partial t} + \boldsymbol{u} \cdot \nabla \rho_0 \right) . \qquad (2.40)$$

(In the last equation γ is also an equilibrium quantity because we have linearized, but for clarity the zero subscript has been omitted.) We see that this is of the same form as Eq. (2.37) but with an additional factor γ.

The adiabatic approximation will generally be a good one when considering dynamical motions of e.g. the deep interiors of stars, where the dynamical timescale is much shorter than the thermal timescale. In that case, differentiating Eq. (2.32) (the linearized equation of motion) with respect to time, and using Eq.(2.40) to eliminate p' and Eq. (2.33) (the linearized continuity equation) to eliminate ρ', yields

$$\frac{\partial^2 \boldsymbol{u}}{\partial t^2} = \frac{1}{\rho_0} \nabla \left[\gamma p_0 \nabla \cdot \boldsymbol{u} + \boldsymbol{u} \cdot \nabla p_0 \right] - \frac{1}{\rho_0^2} \nabla \cdot (\rho_0 \boldsymbol{u}) \nabla p_0 - \nabla \left(\frac{\partial \psi'}{\partial t} \right) . \quad (2.41)$$

In the penultimate term, the equilibrium Eqs. (2.28) have been used to eliminate $\nabla \psi_0$ in favour of ∇p_0.

2.4 Lagrangian Perturbations

We have previously considered perturbations evaluated at a fixed point in space, so for example $p' = p(r, t) - p_0(r)$ is the difference between the actual pressure and the value it would take in equilibrium *at that same point in space*. One can also evaluate perturbations as seen by a fluid element (cf. the material derivative). Such a perturbation will be denoted δp, for example. Now δr is the displacement of a fluid element from the position it would have been at in equilibrium. So

$$\delta p \equiv p(r_0 + \delta r) - p_0(r_0) = p(r_0) + \delta r \cdot \nabla p_0 - p_0(r_0) , \qquad (2.42)$$

where r_0 is the equilibrium position of the fluid element; in the second equation, the first two terms of a Taylor expansion of $p(r_0 + \delta r)$ have been taken: strictly we should have $\delta r \cdot \nabla p$, but $\delta r \cdot \nabla p_0$ is correct up to terms linear in small quantities. Equation (2.42) can be written

$$\delta p(r_0) = p'(r_0) + \delta r \cdot \nabla p_0 \qquad (2.43)$$

where the argument on the left is written r_0 (rather than r) and this is again correct in linear theory. Of course, Eq. (2.43) holds for any quantity, not just pressure. We note that, in linear theory, $\partial \delta f / \partial t = D \delta f / Dt$ (where f is any quantity). The material rate of change of the displacement δr of a fluid element from its equilibrium position is equal to its velocity. Hence

$$u = \frac{D \delta r}{Dt} = \frac{\partial \delta r}{\partial t} . \qquad (2.44)$$

Perturbations such as p' at a fixed point in space are called Eulerian; perturbations such as δp following the fluid are called Lagrangian.

2.5 Sound Waves

Just as the Poisson equation (1.15) has integral solution (1.13), so Eq. (2.34) has solution

$$\psi'(r) = \int \frac{-G \rho'(\tilde{r})}{|r - \tilde{r}|} d\tilde{V} , \qquad (2.45)$$

the integration being over the whole volume of the fluid. In the integral on the right-hand side of (2.45) the positive and negative fluctuations in ρ' tend to cancel out, so that it is often a reasonable approximation to say that $\psi' \approx 0$. Thus we will frequently drop ψ' in Eq. (2.41). Indeed we shall do so in the remainder of this chapter. The ψ' term is of course also absent in problems where self-gravitation is ignored altogether. However it is crucially important in the Jeans instability (see Chapter 10).

Suppose now that we have a *homogeneous* medium, so that equilibrium quantities are independent of position (and hence in particular $\nabla p_0 = 0 = \nabla \psi_0$). Equations (2.32) and (2.33) can then be rewritten

$$\rho_0 \frac{\partial \boldsymbol{u}}{\partial t} = -\nabla p', \qquad \frac{\partial \rho'}{\partial t} = -\rho_0 \nabla \cdot \boldsymbol{u}, \tag{2.46}$$

so taking the divergence of the first of these equations and substituting for $\nabla \cdot \boldsymbol{u}$ from the second gives

$$\frac{\partial^2 \rho'}{\partial t^2} = \nabla^2 p'. \tag{2.47}$$

Suppose further that the perturbations are adiabatic. Now Eq. (2.40) for a homogeneous medium becomes

$$\frac{\partial p'}{\partial t} = c_0^2 \frac{\partial \rho'}{\partial t}, \tag{2.48}$$

where $c_0^2 \equiv \gamma p_0 / \rho_0$ is a constant. Integrating with respect to time gives $p' = c_0^2 \rho'$, which can be used to eliminate p' from Eq. (2.47):

$$\frac{\partial^2 \rho'}{\partial t^2} = c_0^2 \nabla^2 \rho'. \tag{2.49}$$

This is a wave equation (cf. the 1-D analogue $\partial^2 \rho' / \partial t^2 = c_0^2 \partial^2 \rho' / \partial x^2$) and describes sound waves propagating with speed c_0. In fact, c_0 is called the adiabatic sound speed. If we had instead assumed isothermal fluctuations, we would have obtained a wave equation with c_0 replaced by a, the isothermal sound speed; cf. Section 2.1.2.

One can seek plane wave solutions of Eq. (2.49):

$$\rho' = A \exp(i\boldsymbol{k} \cdot \boldsymbol{r} - i\omega t), \tag{2.50}$$

where the amplitude A, frequency ω and wavenumber $\boldsymbol{k}$ are constants. (Here and elsewhere, it should be understood when writing complex quantities that the real part should be taken to get a physically meaningful solution.) Substituting Eq. (2.50) into (2.49), one finds that a non-trivial solution ($A \neq 0$) requires

$$\omega^2 = c_0^2 |\boldsymbol{k}|^2. \tag{2.51}$$

This is known as the *dispersion relation* for the waves. It specifies the relation that must hold between the freqency and wavenumber for the wave to be a solution of the given wave equation. With a suitable choice of phase, one can deduce from (2.50) that

$$\rho' = \alpha \cos(\boldsymbol{k} \cdot \boldsymbol{r} - \omega t),$$

$$p' = \alpha c_0^2 \cos(\boldsymbol{k} \cdot \boldsymbol{r} - \omega t), \tag{2.52}$$

$$\delta\boldsymbol{r} = -\alpha \frac{c_0^2}{\rho_0 \omega^2} \boldsymbol{k} \sin(\boldsymbol{k} \cdot \boldsymbol{r} - \omega t), \tag{2.53}$$

for some constant amplitude α. Note that the adiabatic pressure and density fluctuations are in phase, whereas the displacement $\boldsymbol{r}$ is $\pi/2$ out of phase. A sound wave is called longitudinal, because the fluid displacement is parallel to the wavenumber $\boldsymbol{k}$.

2.6 Surface Gravity Waves

As a second example of a simple wave solution of the linearized perturbed fluid equations, consider incompressible motions ($\nabla \cdot \boldsymbol{u} = 0$) of a fluid of constant density ρ_0 which occupies the region $z < 0$ below the free surface $z = 0$ (so p is constant at the surface). Suppose also that gravity $\boldsymbol{g} = -g\boldsymbol{e}_z$ is uniform and points downwards, and that self-gravity is negligible. This is a reasonable model for ocean waves on deep water, for example. Equation (2.33) implies that $\rho' = 0$. Hence Eq. (2.32) becomes

$$\rho_0 \frac{\partial \boldsymbol{u}}{\partial t} = -\nabla p', \tag{2.54}$$

and taking the divergence of this gives

$$\nabla^2 p' = 0. \tag{2.55}$$

We seek a solution with sinusoidal horizontal variation in the x direction:

$$p'(x, z, t) = f(z) \cos(kx - \omega t) \tag{2.56}$$

(with $k > 0$ for definiteness), where f is an as yet unknown function; and without loss of generality $k > 0$. Substituting this into (2.55) gives

$$\frac{\mathrm{d}^2 f}{\mathrm{d}z^2} = k^2 f \tag{2.57}$$

whence

$$f(z) = A \exp(kz) + B \exp(-kz) . \tag{2.58}$$

The fluid is infinitely deep, and the solution should not become infinite as $z \to -\infty$; hence $B = 0$.

The boundary condition at the free surface is that the pressure at the edge of the fluid should be constant: hence $\delta p = 0$ there. Thus, at the surface,

$$0 = \delta p = p' + \delta \boldsymbol{r} \cdot \nabla p_0 = p' - \rho_0 g \boldsymbol{e}_z \cdot \delta \boldsymbol{r} . \tag{2.59}$$

On the other hand, taking the dot product of Eq. (2.54) with $\boldsymbol{e}_z$, and using Eqs. (2.56) and (2.58) with $B = 0$, yields

$$\boldsymbol{e}_z \cdot \delta \boldsymbol{r} = -\frac{1}{\rho_0 \omega^2} \frac{\partial p'}{\partial z} = -\frac{k}{\rho_0 \omega^2} p' \tag{2.60}$$

everywhere. Hence the boundary condition (2.59) can only be satisfied if ω and k satisfy the dispersion relation

$$\omega^2 = gk . \tag{2.61}$$

It is clear that these are surface waves; for the perturbed quantities all decrease exponentially with depth. In reality, of course, the fluid cannot be infinitely deep, so B is not identically zero. Instead, A and B will have to be chosen such that some boundary condition is satisfied at the bottom of the fluid layer. However, provided the depth of the layer is much greater than k^{-1}, it will generally be the case that B has to be much less than A.

If the layer has depth h, and the condition at $z = -h$ is that the vertical displacement is zero, $B \neq 0$ and by the first part of Eq. (2.60) the lower boundary condition amounts to requiring that $\partial p'/\partial z = 0$ there. All the above equations hold, except the last part of Eq. (2.60). It follows that $B/A = \exp(-kh)$ and the dispersion relation is

$$\omega^2 = gk \tanh kh . \tag{2.62}$$

In the regime $kh \gg 1$ ("deep layer") then this is approximated by Eq. (2.61). In the opposite limit of $kh \ll 1$ ("shallow layer") the dispersion relation approximates to $\omega^2 = (gh)k^2$, i.e. $\omega = \sqrt{gh}\, k$.

2.7   Phase Speed and Group Velocity

Before leaving the topic of waves it is worth noting two different concepts regarding the speed at which waves propagate. Consider a wave which is locally a plane wave, propagating with wavenumber k and with frequency ω. These two quantities are related by a dispersion relation, so $\omega = \omega(k)$. Such a wave is proportional to $e^{ik \cdot x - i\omega(k)t}$. The phase of the wave is $k \cdot x - \omega(k)t$, and wave fronts are surfaces of constant phase. One concept of the speed at which a wave propagates is the *phase speed*. Compare the wave at some location x and time t with the wave at a location $x + n\Delta x$ and slightly different time $t + \Delta t$, where n is a unit vector in any chosen direction. The phase at the second location and time will be the same as at the first location and time if $\Delta x = (\omega / k \cdot n)\Delta t$. Hence we refer to $\omega / (k \cdot n)$ as the phase speed in direction n. In particular, the phase speed in the x-direction is $v_{\mathrm{ph}\,x} = \omega / k_x$, and likewise for the y- and z-directions, provided the wavenumber has a non-zero component in that direction. Sometimes the *phase velocity* is defined to be a quantity with the direction of k and magnitude equal to the phase speed in the direction of k; but it should be noted that the x-, y- and z-components of this 'vector' are not in general the same as the phase speeds in the x-, y- and z-directions. Note also that in directions almost perpendicular to k (so $k \cdot n$ almost zero) the phase speed can become arbitrarily large; but this does not correspond to any physical transport at that speed.

A second concept of the speed of a wave is the *group speed* or *group velocity*. A packet of waves of different wavenumbers but similar to k_0 say propagates physically at a velocity v_g given by $v_{g\,i} = \partial\omega / \partial k_i$, or in shorthand $v_g = \partial\omega / \partial k$, evaluated at $k = k_0$. This is the group velocity. We can also speak of the magnitude of this vector as the group speed. For a proof that this is indeed the velocity at which a wave packet would propagate, see for example the book by Lighthill (1978). This is the velocity at which wave energy propagates.

In the case of pure sound waves, it is straightforward to show from their dispersion relation (2.51) that the group speed is c_0 and that the phase speed normal to the wave fronts is also c_0. Hence in this case these two are equal. In the case of surface gravity waves considered in Section 2.6, the phase speed normal to the wave fronts is ω / k, but differentiating the dispersion relation (2.61) gives that the group speed is only half of this, so in this case the two speeds are not equal. Although in the case of pure sound waves the group velocity is in the direction of the wavenumber, this will not be true for waves in general.

2.8 Order-of-magnitude Estimates for Astrophysical Fluids

2.8.1 *Typical scales*

A system can often be characterized by a typical length scale $\mathcal{L}$, time scale $\mathcal{T}$ and velocity $\mathcal{U}$. These are usually related by $\mathcal{U} = \mathcal{L}/\mathcal{T}$.

The appropriate length scale $\mathcal{L}$ for a particular motion may be different from the size of the whole system – e.g. for sound waves, $\mathcal{L}$ might be the wavelength, $\mathcal{T}$ the period and $\mathcal{U}$ the sound speed.

For example, for motion in a gravitational field, with length scale $\mathcal{L}$, the time scale is $\mathcal{T} \sim (\mathcal{L}/g)^{1/2}$. For motion in a star's gravitational field with $g = GM/R^2$, where $R = \mathcal{L}$ is the radius of the star and M its mass,

$$\mathcal{T} \sim t_{\text{dyn}} = \left(\frac{R^3}{GM}\right)^{1/2} \propto (\text{mean density})^{-1/2} . \tag{2.63}$$

This is the typical time scale for oscillations of e.g. Cepheid variable stars. t_{dyn} is called the dynamical timescale.

2.8.2 *Importance of viscosity*

Molecular viscosity, which provides tangential forces in fluids, comes about microscopically because molecules from faster-flowing fluid diffuse into slower-flowing fluid, and vice versa. As can be seen from e.g. Eq. (1.9), the viscosity μ has dimensions $ML^{-1}T^{-1}$ where M, L, T denote mass, length and time respectively. Let the molecules have mean velocity v and mean free path l. Then on dimensional grounds,

$$\mu \sim \rho v l . \tag{2.64}$$

Often people work with the *kinematic viscosity* $\nu \equiv \mu/\rho$; thus $\nu \sim vl$. Equation (2.64) can also instructively be deduced by considering the tangential force at a plane interface between two fluids moving at different speeds, assuming that such force comes about by molecules diffusing a distance of order l across the boundary at a speed v and depositing their momentum in the new environment, noting the relation that force is equal to rate of change of momentum.

To make further progress, we need to relate v and l to macroscopic properties of the fluid. If the collisional cross-section for the molecules is

σ (not to be confused with the stress tensor!), and the number density of particles is n (so that $\rho = nm$, where m is the mean molecular mass), then on average between collisions a particle sweeps out a cylinder of volume σl and thus such a cylinder must contain on average one particle: $nl\sigma \sim 1$. Hence $l \sim m/\rho\sigma$. The mean kinetic energy of a molecule $\sim k_BT$, where k_B is Boltzmann's constant, so

$$v \sim \left(\frac{k_BT}{m}\right)^{1/2}. \qquad (2.65)$$

Thus

$$\mu \sim \frac{(k_Bm)^{1/2}}{\sigma}T^{1/2}. \qquad (2.66)$$

For a crude estimate, $\sigma \sim$ (radius of atom)^2, so about 10^{-20}m^2. (This underestimates σ in an ionized gas, where electromagnetic interactions are important.) The mean mass $m = \tilde{\mu}m_u$, where $\tilde{\mu}$ here denotes the mean molecular weight and m_u is the atomic mass unit. Assuming reasonably that all the constants implied in the '$\sim$' relations above are of order unity, this gives

$$\mu \simeq 10^{-5}(\tilde{\mu}T)^{1/2}\,\text{kg}\,\text{m}^{-1}\,\text{s}^{-1} \qquad \nu \simeq 10^{-5}\frac{(\tilde{\mu}T)^{1/2}}{\rho}\,\text{m}^2\,\text{s}^{-1}. \qquad (2.67)$$

where T is in Kelvin and ρ is in $\text{kg}\,\text{m}^{-3}$. (See Appendix A for the values of physical constants.)

Now the left-hand side of the Navier-Stokes equation (1.9) is $\rho D\boldsymbol{u}/Dt$, while a typical viscous term is $\mu\nabla^2\boldsymbol{u}$. If viscosity were dominant, then the timescale of motion would be determined by its effect:

$$\rho\frac{\mathcal{U}}{\mathcal{T}_\nu} \sim \mu\frac{\mathcal{U}}{\mathcal{L}^2} \qquad (2.68)$$

or, rearranging, $\mathcal{T}_\nu \sim \mathcal{L}^2/\nu$. For typical stellar values ($T = 10^6\,\text{K}$, $\rho = 1\,\text{kg}\,\text{m}^{-3}$, $\tilde{\mu} = 1$, $\mathcal{L} = 10^8\,\text{m}$) we deduce using (2.67) that the viscous timescale is of order $10^{21}\,\text{s} \simeq 3 \times 10^{13}$ years. Even for a star this is a very long time, so molecular viscosity is unlikely to be important on stellar scales. This will generally be true for astrophysical fluids, though some form of viscosity is important in e.g. accretion disks (see Chapter 9).

A commonly used measure of the importance of viscous effects is the Reynolds number Re, which is the ratio of the advection term (implicit) on the left-hand side of Eq. (1.9) to the viscous term on the right-hand side:

$$\frac{|\rho \boldsymbol{u} \cdot \nabla \boldsymbol{u}|}{|\mu \nabla^2 \boldsymbol{u}|} \sim \frac{\rho \mathcal{U}^2/\mathcal{L}}{\mu \mathcal{U}/\mathcal{L}^2} \sim \frac{\mathcal{U}\mathcal{L}}{\nu} \equiv \mathrm{Re} \,. \tag{2.69}$$

Viscous effects are important if $\mathrm{Re} \lesssim 1$. For stellar scales, for speeds close to the sound speed inside the Sun ($\simeq 10^5 \,\mathrm{m\,s^{-1}}$), $\mathrm{Re} \simeq 10^{18}$, and so even for substantially subsonic speeds the Reynolds number is generally very much greater than unity. This shows once again that molecular visccous effects are generally negligible in the stellar context. However, small-scale turbulent flows can have an effect on the mean large-scale motion similar to that of viscosity: this is known as turbulent viscosity. It may well be the source of "viscosity" in many astrophysical viscous accretion disks, for example.

2.8.3 *The adiabatic approximation*

Suppose $\mathcal{T}_F$ is the timescale for the transfer of heat (by flux $\boldsymbol{F}$). If this is much greater than the timescale of the motion then one can treat the motion as adiabatic ($\delta Q = 0$). This is the adiabatic approximation. For the Sun, $\mathcal{T}_F \approx 10^7$ years in the interior, and about one day near the surface. The fundamental period of oscillation of the Sun is about one hour (see Chapter 11), so for most purposes the adiabatic approximation is excellent for describing oscillations of the Sun. In the solar atmosphere, however, $\mathcal{T}_F$ can be much shorter, in fact so short that there are some circumstances in which one can treat the motion as isothermal (Section 2.3).

2.8.4 *The approximation of incompressibility*

The flow is incompressible if $D\rho/Dt = 0$ for then the density of a fluid element does not change with time. By the continuity equation (2.2) this is equivalent to $\nabla \cdot \boldsymbol{u} = 0$. (Some authors prefer to take $\nabla \cdot \boldsymbol{u} = 0$ as the definition of incompressibility, and $D\rho/Dt = 0$ as the consequence of that.) Roughly speaking, the conditions for this to hold are that $\mathcal{U}$ is much less than the sound speed and that $\mathcal{L}$ is much smaller than the pressure scale height

$$H_p = p \left| \frac{\mathrm{d}p}{\mathrm{d}z} \right|^{-1} . \tag{2.70}$$

For example, for the Earth's atmosphere $H_p \simeq 10\,\mathrm{km}$. Of course, compressibility cannot be ignored for modelling sound waves!

Chapter 3

Theory of Rotating Bodies

Most if not all objects in the universe rotate, and the effects of rotation are important to an understanding of the structure and dynamics of many astrophysical systems. Rotation is indeed sufficiently important to the subject of astrophysical fluid dynamics that we return to it several times in addition to the present chapter: in Chapter 4 on fluid instabilities, in Chapter 7 on the dynamics of planetary atmospheres, in Chapter 9 on accretion disks, and elsewhere. The present chapter establishes the equations of motion in a rotating frame of reference and considers the equilibrium structure and shape of a slowly and uniformly rotating star (or gaseous planet). We shall also consider briefly the internal dynamics of a rotating star, and some consequences of orbital rotation of stars in a binary system.

A great deal of research has been made into equilibria of rotating bodies, particularly in the case of bodies with uniform density, by such illustrious names as Laplace, Jacobi, Liouville, Riemann, Poincaré, Lord Kelvin and Jeans. Much interesting detail of the results and history can be found in e.g. Lyttleton (1953), Lebovitz (1967), Chandrasekhar (1969) and Tassoul (1978). Just to give some brief historical context, we mention that in the case of bodies of uniform rotation there are two families of equilibrium configurations. One consists of the Maclaurin spheroids: these are axisymmetric configurations. The second family consists of the Jacobi ellipsoids, which are non-axisymmetric. When the rotation is sufficiently fast, as measured by the quantity $\Omega^2/2\pi G\bar{\rho}$ where $\bar{\rho}$ is the density, then the Maclaurin sequence terminates and for faster rotation the only equilibrium configurations for homogeneous bodies are the tri-axial Jacobi ellipsoids.

3.1 Equation of Motion in a Rotating Frame

When considering rotating systems, it is usually most convenient to work in a rotating frame of reference. The fluid velocity u is the rate of change of a fluid element's position r with time. If we use D/Dt to denote rate of change as measured in an inertial (nonrotating) frame, and d/dt to denote rate of change as measured in the rotating frame, then

$$\frac{D\boldsymbol{r}}{Dt} = \frac{d\boldsymbol{r}}{dt} + \boldsymbol{\Omega}\times\boldsymbol{r}\,, \qquad (3.1)$$

where Ω is the angular velocity of the rotating frame relative to the inertial one. Applying the same rule a second time gives

$$\frac{D^2\boldsymbol{r}}{Dt^2} = \left(\frac{d}{dt} + \boldsymbol{\Omega}\times\right)^2 \boldsymbol{r} = \frac{d^2\boldsymbol{r}}{dt^2} + 2\boldsymbol{\Omega}\times\frac{d\boldsymbol{r}}{dt} + \boldsymbol{\Omega}\times(\boldsymbol{\Omega}\times\boldsymbol{r})\,, \qquad (3.2)$$

where in the last step we have now assumed that the rotation rate Ω does not vary with time. There are some subtleties to vectors in rotating frames and calculating their rates of change, and the reader who would like more details is referred to Chapter 3 of Jeffreys & Jeffreys (1956).

Now in the inertial frame, Eq. (2.1) is the equation of motion; so substituting for $Du/Dt \equiv D^2r/Dt^2$ from Eq. (3.2), and identifying dr/dt as the velocity as measured in the rotating frame, gives the following equation of motion in the rotating frame:

$$\frac{d\boldsymbol{u}}{dt} = -\frac{1}{\rho}\nabla p - \nabla\psi - 2\boldsymbol{\Omega}\times\boldsymbol{u} - \boldsymbol{\Omega}\times(\boldsymbol{\Omega}\times\boldsymbol{r})\,. \qquad (3.3)$$

The last term is the centrifugal acceleration; the penultimate term is the Coriolis acceleration, which is zero if $u = 0$ and is perpendicular to the velocity otherwise.

3.2 Equilibrium Equations for a Slowly Rotating Body

In this chapter we shall consider how to calculate the shape of a fluid body that is rotating slowly with a uniform rotation rate. We shall consider in particular the case of a slowly rotating star; but the equations apply equally well to, for example, a slowly rotating gaseous planet. It will be assumed that in the absence of rotation the body would be spherically symmetric, and that rotation induces a weak distortion of the shape from spherical symmetry. For slow rotation, the distorted body is axisymmetric about the rotation axis, as one would expect. Although we shall not consider it here, faster rotation can give rise to some surprises, notably the Jacobi ellipsoids which are triaxial figures of equilibrium. For a fuller exposition of the subject, see the classic texts by Chandrasekhar (1969) and Lyttleton (1953).

We work in a frame rotating with the body: in that frame the equilibrium is described by $u = 0$ and $\partial/\partial t = 0$. Since we are modelling a nearly spherical configuration, we use spherical polar coordinates (r, θ, ϕ). In these coordinates, writing $\mathbf{\Omega}$ as Ωe_z (where e_z is a unit vector along the polar axis,

$$\mathbf{\Omega} \times r = \Omega r \sin\theta e_\phi . \tag{3.4}$$

This is just the velocity of the fluid as seen from the nonrotating frame, if it is at rest in the rotating frame. Moreover,

$$-\mathbf{\Omega} \times (\mathbf{\Omega} \times r) = \Omega^2 r \sin^2\theta \, e_r + \Omega^2 r \sin\theta \cos\theta \, e_\theta$$
$$= \nabla \left(\frac{1}{2} \Omega^2 r^2 \sin^2\theta \right) . \tag{3.5}$$

So the centrifugal acceleration can be written as the gradient of a potential. Note that $r \sin\theta$ is simply the distance from the rotation axis. Indeed the above would be simpler in cylindrical polar coordinates (ϖ, ϕ, z) since coordinate ϖ (pronounced "pomega") is the distance from the axis: Eq. (3.5) would become

$$-\mathbf{\Omega} \times (\mathbf{\Omega} \times r) = \nabla \left(\frac{1}{2} \Omega^2 \varpi^2 \right) . \tag{3.6}$$

We are interested in finding equilibrium solutions of Eq. (3.3). Setting $u = 0$ gives

$$\nabla p = -\rho \nabla \Phi , \tag{3.7}$$

where

$$\Phi = \psi - \frac{1}{2} \Omega^2 r^2 \sin^2\theta \tag{3.8}$$

is the total effective gravitational potential (gravitational plus centrifugal).

We can argue qualitatively from Eq. (3.5) what the effect of rotation on the equilibrium shape of the body will be. At the poles ($\theta = 0, \pi$) the centrifugal acceleration $-\mathbf{\Omega} \times (\mathbf{\Omega} \times r)$ is zero, and it is radially outwards at the equator ($\theta = \pi/2$). It thus reduces the effective gravitational acceleration at the equator, i.e. the centreward pull is not so strong there as at the poles and so instead of being spherical the body "bulges" at the equator.

The gradient vector ∇f of any scalar is perpendicular to surfaces of constant f, so a normal n to the surface of constant f satisfies $n \times \nabla f = 0$. It follows from Eq. (3.7) that surfaces of constant p are also surfaces of constant Φ, and vice versa. Thus we can write $p = p(\Phi)$, and so

$$\nabla p = \frac{\mathrm{d}p}{\mathrm{d}\Phi} \nabla \Phi . \tag{3.9}$$

Substituting this into Eq. (3.7) yields

$$\rho = -\frac{\mathrm{d}p}{\mathrm{d}\Phi} , \tag{3.10}$$

so ρ is also a function of Φ, i.e. $\rho = \rho(\Phi)$.

Henceforward, for definiteness, we shall speak of the body as being a star, but it could equally be a gaseous planet, for example. The outer surface of the star is a surface of constant pressure (because the pressure outside is constant, say zero) and so Φ is constant on the surface.

We consider the case where if the star were not rotating it would be spherically symmetric, and rotation induces a *weak* distortion from sphericity. We suppose that the star has mass M and (in the nonrotating case) radius R.

3.3   The Roche Model

On the surface, $\Phi = \psi - \frac{1}{2}\Omega^2 r^2 \sin^2 \theta$ is constant. Let us approximate the gravitational potential ψ by what it would be in the nonrotating case:

$$\psi = -\frac{GM}{r} \tag{3.11}$$

at the surface and outside the star. This is equivalent to approximating the gravitational potential as if all the mass were at the centre and is called the *Roche model*. This is a reasonable approximation in the case of a centrally condensed star, in which most of the mass is concentrated near the centre.

We suppose that the surface of the rotating star is described by

$$r = R(1 + \epsilon(\theta)) , \tag{3.12}$$

where $\epsilon(\theta)$ is a function of θ. Then

$$\Phi_{\text{surface}} = -\frac{GM}{R(1 + \epsilon(\theta))} - \frac{1}{2}\Omega^2 R^2 (1 + \epsilon(\theta))^2 \sin^2 \theta \tag{3.13}$$

is constant (i.e. independent of θ). The rotation is slow and the distortion weak, so Ω^2 and ϵ are small and we neglect products of small quantities. Then (3.13) implies that

$$-\frac{GM}{R}(1 - \epsilon(\theta)) - \frac{1}{2}\Omega^2 R^2 \sin^2 \theta \tag{3.14}$$

is independent of θ, i.e.

$$\epsilon(\theta) = \frac{1}{2}\frac{\Omega^2 R^3}{GM} \sin^2 \theta + \text{constant} . \tag{3.15}$$

Note that $\Omega^2 R$ is the equatorial acceleration due to centrifugal forces; and GM/R^2 is the gravitational acceleration. So the dimensionless quantity $\Omega^2 R^3/(GM)$ is the ratio of centrifugal acceleration to gravitational acceleration.

The radii at the pole and at the equator are obtained from Eq. (3.12) by putting $\theta = 0$ and $\theta = \pi/2$ respectively. Thus the relative difference between equatorial and polar radii is

$$\frac{R(1 + \epsilon(\pi/2)) - R(1 + \epsilon(0))}{R} = \frac{1}{2}\frac{\Omega^2 R^3}{GM} . \tag{3.16}$$

Thus the relative difference in radii, which is a measure of the shape distortion, is $\Omega^2 R^3/(GM)$ times a coefficient of order unity.

The only thing wrong with this argument is the use of Eq. (3.11) to describe the gravitational potential. We should properly use the gravitational potential appropriate to the distorted star. We proceed to do this now.

3.4   Chandrasekhar-Milne Expansion

The missing ingredient in the previous section was a proper treatment of the gravitational potential of the distorted star. In the Chandrasekhar-Milne expansion, one considers the $O(\Omega^2)$ perturbation not only to the shape of the star but also to its gravitational potential. The procedure is described in more detail in Tassoul (1978).

We know that on the surface Φ is constant; also $\rho = \rho(\Phi)$ everywhere. Now the gravitational potential satisfies Poisson's equation (1.15); and it is straightforward to demonstrate that

$$\nabla^2 \left(-\frac{1}{2}\Omega^2 r^2 \sin^2\theta \right) = -2\Omega^2 . \qquad (3.17)$$

It follows then from the definition (3.8) of Φ that

$$\nabla^2\Phi = 4\pi G\rho - 2\Omega^2 . \qquad (3.18)$$

The problem involves the ∇^2 operator, so it is more natural to write the θ-dependence not as $\sin^2\theta$ but in terms of Legendre polynomials of $\cos\theta$: $P_n(\cos\theta)$, since

$$V = r^n P_n(\cos\theta) \quad \text{and} \quad V = r^{-(n+1)} P_n(\cos\theta) \qquad (3.19)$$

are solutions of Laplace's equation, $\nabla^2 V = 0$. The first three Legendre polynomials are

$$P_0(x) = 1, \quad P_1(x) = x, \quad P_2(x) = \frac{1}{2}(3x^2 - 1). \qquad (3.20)$$

Thus Φ may be rewritten as

$$\Phi = \psi - \frac{1}{3}\Omega^2 r^2 \left(1 - P_2(\cos\theta)\right) . \qquad (3.21)$$

Let us write

$$\Phi = \psi_u(r) + \Phi'(r,\theta) , \qquad \rho = \rho_u(r) + \rho'(r,\theta) , \qquad (3.22)$$

where ψ_u and ρ_u are the gravitational potential and density in the spherically symmetric, nonrotating star, and the primed quantities are small perturbations, of order $\Omega^2 R^3/(GM)$, induced by the rotation. As before, we express the stellar surface as $r = R(1+\epsilon(\theta))$, where ϵ is the same order as the other small perturbations; and we neglect products of small quantities. Recalling that $\rho = \rho(\Phi)$,

$$\rho_u + \rho' = \rho(\psi_u + \Phi') = \rho(\psi_u) + \Phi' \left.\frac{d\rho}{d\Phi}\right|_{\psi_u} . \qquad (3.23)$$

Matching zero-order terms gives $\rho_u = \rho(\psi_u)$, and first-order terms give

$$\rho' = \Phi' \left.\frac{d\rho}{d\Phi}\right|_{\psi_u} . \qquad (3.24)$$

Hence Eq. (3.18) becomes

$$\nabla^2(\psi_u + \Phi') = 4\pi G \left(\rho_u + \Phi' \left.\frac{d\rho}{d\Phi}\right|_{\psi_u} \right) - 2\Omega^2 . \tag{3.25}$$

The zero-order terms give $\nabla^2\psi_u = 4\pi G\rho_u$, while the first-order terms yield

$$\nabla^2\Phi' = 4\pi G \left.\frac{d\rho}{d\Phi}\right|_{\psi_u} \Phi' - 2\Omega^2 . \tag{3.26}$$

Also, Φ is constant on the surface, so

$$\psi_u\left(R(1+\epsilon)\right) + \Phi'(r,\theta) \tag{3.27}$$

is independent of θ; which, after expanding ψ_u as the first two terms of a Taylor series expansion, gives

$$\epsilon(\theta) = -\frac{1}{R} \left(\left.\frac{d\psi_u}{dr}\right|_R \right)^{-1} \Phi'(R,\theta) \ (+ \text{ constant}) . \tag{3.28}$$

Equation (3.26) is an inhomogeneous differential equation for Φ': once Φ' is found, the shape of the surface follows from Eq. (3.28).

One condition on Φ' is that the solution must be regular in the interior, and in particular at $r = 0$. To find a second condition, we must ensure that the gravitational potential ψ matches smoothly onto the external gravitational field.

Outside the star,

$$\psi = -\frac{GM}{r} + \sum_{n=0}^{\infty} A_n \left(\frac{R}{r}\right)^{n+1} P_n(\cos\theta) \tag{3.29}$$

since $\nabla^2\psi = 0$ there by Eq. (1.15), where A_n is of order $\Omega^2 R^3/GM$.

Inside the star,

$$\psi = \Phi + \frac{1}{3}\Omega^2 r^2 \left\{1 - P_2(\cos(\theta)\right\} = \psi_u(r) + \Phi' + \frac{1}{3}\Omega^2 r^2 \left\{1 - P_2(\cos(\theta)\right\} . \tag{3.30}$$

To allow complete generality, we would now write

$$\Phi'(r,\theta) = \sum_{n=0}^{\infty} \Phi_n(r) P_n(\cos\theta), \quad \epsilon(r,\theta) = \sum_{n=0}^{\infty} \epsilon_n(r) P_n(\cos\theta) \tag{3.31}$$

(see Tassoul 1978). However, to avoid needless algebra, we note that the problem for uniform rotation has only $P_2(\cos\theta)$ and $P_0(\cos\theta)$ angular dependence — see Eqs. (3.26) and (3.30) — and so we anticipate the solution to be

$$\Phi'(r,\theta) = \Phi_0(r) + \Phi_2(r)P_2(\cos\theta), \quad \epsilon(r,\theta) = \epsilon_0(r) + \epsilon_2(r)P_2(\cos\theta),$$
(3.32)

and similarly for the external field (3.29). The surface boundary condition becomes that of requiring ψ and $\partial\psi/\partial r$ to be continuous there: for otherwise, since ψ satisfies the Poisson equation (1.15), a discontinuity in one of these quantities would imply that there was an infinite density at the surface. Continuity of ψ means, equating (3.29) and (3.30), that

$$\psi_u\left(R(1 + \epsilon_0 + \epsilon_2 P_2)\right) + \frac{1}{3}\Omega^2 R^2(1 - P_2) + \Phi_0(R) + \Phi_2(R)P_2$$
$$= \frac{-GM}{R(1 + \epsilon_0 + \epsilon_2 P_2)} + A_0 + A_2 P_2,$$
(3.33)

correct to $O(\Omega^2)$; i.e.

$$\psi_u(R) + R(\epsilon_0 + \epsilon_2 P_2)\frac{d\psi_u}{dr}(R) + \frac{1}{3}\Omega^2 R^2(1 - P_2) + \Phi_0(R) + \Phi_2(R)P_2$$
$$= -\frac{GM}{R} + \frac{GM}{R}(\epsilon_0 + \epsilon_2 P_2) + A_0 + A_2 P_2.$$
(3.34)

Similarly, continuity of $\partial\psi/\partial r$ implies

$$\frac{d\psi_u}{dr}(R) + R(\epsilon_0 + \epsilon_2 P_2)\frac{d^2\psi_u}{dr^2}(R) + \frac{2}{3}\Omega^2 R(1 - P_2) + \left.\frac{d\Phi_0}{dr}\right|_R + \left.\frac{d\Phi_2}{dr}\right|_R P_2$$
$$= \frac{GM}{R^2} - 2\frac{GM}{R^2}(\epsilon_0 + \epsilon_2 P_2) - \frac{A_0}{R} - 3\frac{A_2}{R}P_2.$$
(3.35)

To zero-order, these two equations give simply

$$\psi_u(R) = -\frac{GM}{R}, \quad \frac{d\psi_u}{dr}(R) = \frac{GM}{R^2}.$$
(3.36)

The first-order terms proportional to $P_2(\cos\theta)$ give two equations which, after A_2 has been eliminated between them, yields

$$\frac{d\Phi_2}{dr} + \frac{3}{R}\Phi_2 + \left(\frac{d^2\psi_u}{dr^2} + \frac{2GM}{R^3}\right)R\epsilon_2 = \frac{5}{3}\Omega^2 R,$$
(3.37)

where all functions are evaluated at $r = R$. Now recalling that ψ_u obeys Poisson's equation so

$$\frac{d^2\psi_u}{dr^2} + \frac{2}{r}\frac{d\psi_u}{dr} = 4\pi G\rho_u , \tag{3.38}$$

and using Eq. (3.28) with (3.32) to eliminate ϵ_2, Eq. (3.37) becomes

$$\frac{d\Phi_2}{dr} + \frac{3}{R}\Phi_2 - \frac{4\pi R^2}{M}\rho_u\Phi_2 = \frac{5}{3}\Omega^2 R \tag{3.39}$$

at $r = R$.

Equation (3.39) provides the surface boundary condition that must be applied to the differential equation for Φ_2. Taking just the $P_2(\cos\theta)$ terms from Eq. (3.26), this differential equation is

$$\nabla^2(\Phi_2 P_2) = 4\pi G \left.\frac{d\rho}{d\Phi}\right|_u \Phi_2 P_2 . \tag{3.40}$$

Now $\nabla^2 P_2 = -6P_2/r^2$ and

$$\left.\frac{d\rho}{d\Phi}\right|_u = \frac{d\rho_u}{dr}\bigg/\frac{d\psi_u}{dr} = -\frac{r^2}{Gm}\frac{d\rho_u}{dr} ; \tag{3.41}$$

so Eq. (3.40) gives the following ordinary differential equation for Φ_2:

$$\frac{1}{r^2}\frac{d}{dr}\left(r^2\frac{d\Phi_2}{dr}\right) - \frac{6}{r^2}\Phi_2 = -\frac{4\pi r^2}{m(r)}\frac{d\rho_u}{dr}\Phi_2 , \tag{3.42}$$

with boundary condition (3.39) at $r = R$ and Φ_2 regular at $r = 0$.

In the general case it would be necessary to solve the above equation numerically. However, in the special case where ρ_u is constant, it is straightforward to find the solutions of Eq. (3.42) in the form $\Phi_2 = Ar^p$ for constants A and p, and after applying boundary conditions to deduce that $\Phi_2 = (5/6)\Omega^2 r^2$. Further, it follows from Eq. (3.28) that $\epsilon_2 = -(5/6)\Omega^2 R^3/GM$. Hence the difference between the equatorial and polar radii, divided by R, is $(5/4)\Omega^2 R^3/GM$ in the case of a homogeneous stellar model.

3.5 Dynamics of Rotating Stellar Models

We have not so far considered how energy is transported in the rotating star. A well-known result, which is discussed at length by Tassoul (1978), is that one cannot have a uniformly rotating star in strict radiative equilibrium.

Assuming the contrary leads to what is known as *von Zeipel's paradox*. The same is true if the rotation rate is a function only of distance from the rotation axis. We conclude therefore that the rotation rate must have a more general form, depending on cylindrical polar coordinate z as well as distance from the axis, or that strict radiative equilibrium does not hold.

We consider now the latter possibility. The von Zeipel paradox in effect says that the radiative flux cannot be balanced everywhere by the energy generation. Some regions have a net influx of heat: these will heat up and tend to rise under buoyancy. Others will cool and sink. This tends to set up motions in meridional planes: this is called meridional circulation. It can be shown, e.g. Kippenhahn & Weigert (1990) and Tassoul (1978), that the global timescale for mixing by the meridional circulation, known as the Eddington-Sweet timescale, is of order τ_{KH}/χ where $\tau_{KH} \equiv GM^2/RL$ is the Kelvin-Helmholtz timescale (L being the luminosity) and $\chi \simeq \Omega^2 R^3/GM$.

For the Sun, τ_{KH} is about 10^7 years, and $\chi \simeq 10^{-5}$; so the Eddington-Sweet timescale is about 10^{12} years, much longer than the Sun's age. *Local* circulation timescales can be much shorter, however.

For the Sun, the Eddington-Sweet timescale is much greater even than the nuclear timescale, but this is not so for some more massive stars. Yet the observational evidence does not support the idea that these stars are mixed, as these timescales would suggest. The explanation (cf. Kippenhahn & Weigert 1990) is that mixing is opposed and stopped by composition gradients (and hence gradients in the mean molecular weight).

It should be mentioned that, although one can postulate some arbitrary rotation profile for the interior of a star, this will not necessarily be stable, and hence will not necessarily be realizable in a real star. An example of a stability consideration is the *Rayleigh criterion* (see Section 4.3).

3.6   Solar Rotation

Detailed observations have been carried out of the Sun's surface rotation rate over many years, by tracking surface features such as sunspots and, more recently, using Doppler velocity measurements. The rotation rate varies with latitude, the equatorial regions having a rotation period of about 25 days while high latitudes rotate more slowly, with periods in excess of one month. The surface rotation rate is commonly expressed in an expansion in $\cos^2 \theta$, where θ is co-latitude, e.g. Snodgrass (1983). The Doppler rate,

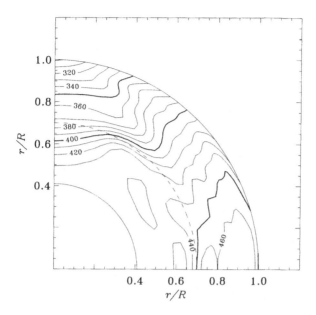

Fig. 3.1 Rotation rate inside the Sun, as inferred by helioseismology. The contours are labelled in nHz and have a spacing of 10 nHz. The horizontal axis is in the Sun's equatorial plane, the vertical axis is the axis of rotation. Distances are in units of the Sun's photospheric radius. The dashed line marks the base of the Sun's convection zone.

for example, is given by Ulrich *et al.* (1988) as approximately

$$\Omega/2\pi \;=\; \left(451.5 \;-\; 65.3 \cos^2 \theta \;-\; 66.7 \cos^4 \theta\right) \text{nHz} . \tag{3.43}$$

The rotation rates determined from tracking different features do not agree with the above rate or with each other precisely. Sunspots, for example, rotate at 10–15 nHz faster than (3.43) indicates at low latitudes. Their movement may be more indicative of the rotation rate in a somewhat deeper subsurface region where the spots may be rooted.

For at least a century the surface rate has not changed by more than 5 per cent. However, variations of the order of 1% have been detected as zonal bands of flow migrating from mid-latitudes to the equator with a period of about 11 years. These are called torsional oscillations, though this is a bit of a misnomer.

The rotation rate of the interior itself has been revealed in the past two decades by helioseismic imaging using observed frequencies of global acoustic modes of the Sun (see Section 12.9). Waves propagating in the same direction as the rotation have a slightly higher frequency than those propagating against the rotation, and the difference in frequency depends on the rotation rate. The results of such imaging in the outer 60 per cent or so of the solar interior are shown in Fig. 3.1. The outer 30 per cent of the solar interior is the convectively unstable convection zone. In this region, the differential rotation with latitude is similar to that seen at the surface, so the contours of constant rotation are nearly radial. Only at low latitudes do we see something like Taylor columns (see Section 7.4). By contrast the radiative interior beneath the convection zone appears to be in a state of nearly rigid-body rotation, to the extent that it can be measured at present using helioseismology. Between the two regions is a layer of strong shear, called the tachocline. There is also strong radial shear in the region just beneath the surface. The helioseismic findings and their theoretical interpretation are reviewed by Thompson *et al.* (2003).

Young stars are observed to rotate much faster than the Sun, and it is believed that stars lose angular momentum from their surface layers through stellar winds. This loss is only communicated to the stellar interior if there are ways to redistribute the angular momentum inside the star. As we shall see in Chapter 4, shear in a flow induces instabilities (Sections 4.3, 4.4), so a sufficiently steep rotational gradient would become unstable. Turbulence might then transport the angular momentum. Magnetic fields, via the action of Alfvén waves, may also redistribute angular momentum. A weak magnetic field may indeed be responsible for the nearly rigid rotation of much of the radiative interior and may also stop the spread of the tachocline gradient further down into the Sun.

Including magnetic Lorentz force $j \times B$ and a viscous term $\mathcal{D}$, the momentum equation in a frame rotating with steady angular velocity Ω_0 (which we take to be the mean solar rotation rate) is

$$\rho \frac{\partial u}{\partial t} = -\rho(u \cdot \nabla)u - \nabla p + \rho \nabla \Phi - 2\rho \Omega_0 \times u + \frac{1}{\mu_0} j \times B + \mathcal{D} . \quad (3.44)$$

As usual, u is the residual velocity in the rotating frame. The total angular rotation rate is

$$\Omega(r) = \Omega_0 + \frac{\langle u_\phi \rangle}{r \sin \theta} ; \quad (3.45)$$

$\langle \ldots \rangle$ denotes an average over longitude. An equation for the conservation of angular momentum density $J = \rho(r \sin \theta)^2 \Omega$ can be obtained by multiplying the ϕ-component of eq. (3.44) describing the rate of change of momentum by the distance $r \sin\theta$ from the rotation axis, and using the continuity equation (1.3):

$$\frac{\partial J}{\partial t} = -\nabla \cdot (\mathcal{F}_{\mathrm{MC}} + \mathcal{F}_{\mathrm{RS}} + \mathcal{F}_{\mathrm{EM}} + \mathcal{F}_{\mathrm{V}}) \, . \tag{3.46}$$

Here the term on the right comprises (minus) the divergence of various angular-momentum fluxes. The first two are

$$\mathcal{F}_{\mathrm{MC}} = (\langle u_r \rangle \boldsymbol{e}_r + \langle u_\theta \rangle \boldsymbol{e}_\theta) \, J \, , \tag{3.47}$$

the flux of angular momentum due to the meridional circulation in the (r,θ)-directions, and

$$\mathcal{F}_{\mathrm{RS}} = r \sin\theta \left(\langle \rho u'_r u'_\phi \rangle \boldsymbol{e}_r + \langle \rho u'_\theta u'_\phi \rangle \boldsymbol{e}_\theta \right) \, , \tag{3.48}$$

the Reynolds stress term that arises from non-zero correlations between turbulent fluctuations $\boldsymbol{u}' \equiv \boldsymbol{u} - \langle \boldsymbol{u} \rangle$ in the velocity in the ϕ-direction and the other two directions. The remaining terms represent the transport due to electromagnetic Maxwell stresses

$$\mathcal{F}_{\mathrm{EM}} = \frac{r \sin\theta}{\mu_0} \left(\langle \rho B_r B_\phi \rangle \boldsymbol{e}_r + \langle \rho B_\theta B_\phi \rangle \boldsymbol{e}_\theta \right) \tag{3.49}$$

and viscous diffusion

$$\mathcal{F}_{\mathrm{V}} = -\nu \rho (r \sin\theta)^2 \nabla\Omega \tag{3.50}$$

respectively (e.g. Thompson *et al.* 2003). In deriving Eqs. (3.47)–(3.50) we have neglected longitudinal variations in ρ and ν.

Viscous forces are presumably negligible in the solar interior, and the Maxwell stresses are also likely small in the bulk of the convection zone (though possibly not in the radiative interior and tachocline, nor in sunspots). Hence a steady-state rotation in the convection zone indicates a balance between the divergences of the fluxes of angular momentum caused by meridional circulation and Reynolds stresses due to turbulence. If the Rossby number is small and the flow barotropic, then the Taylor-Proudman theorem (Section 7.4) states that the rotation rate will be constant on cylindrical surfaces aligned with the rotation axis. This is evidently not the case in the solar convection zone (Fig. 3.1) except perhaps at low latitudes. This is at least partly due to baroclinicity ($\nabla p \times \nabla \rho \neq 0$) driving a meridional circulation which redistributes angular momentum. Latitudinal variations in heat transport due to rotational modulation of the turbulence cause a thermal wind (Section 7.4); but also the Rossby number is likely not small for *some* scales of motion in the turbulent convection zone, breaking the conditions for the Taylor-Proudman theorem to apply.

It is possible to model the rotation in the convection zone using mean-field models (cf. Section 5.3.2) but knowing how to prescribe the Reynolds stresses from the mean-field velocity is a difficulty with this approach. Recent large-scale numerical simulations (e.g. large-eddy simulations) capturing some of the turbulent nature of the convection zone, can produce rotation profiles that are qualitatively similar to what is being found in helioseismology (see Thompson *et al.* 2003 for a review).

3.7 Binary Stars

Many stars are found to be in binary systems. The orbits of stars in a close binary system tend to become circular over time, due to tidal forces. Consider a binary system in which the two components are in circular orbits about their common centre of mass O, and in which the two stars corotate so as to always show the same side to the other star. In this system there is a rotating frame in which the stars are completely stationary. If Ω is the angular velocity of each star about O, in an inertial frame, then of course Ω is also the angular velocity of the rotating frame.

Suppose that the separation distance between the two stars is a, that their masses are M_1, M_2, and that their respective distances from O are μa and $(1 - \mu)a$. Since O is the centre of mass,

$$\mu = \frac{M_2}{M_1 + M_2}. \tag{3.51}$$

Also the gravitational force on star 1 towards star 2 (and hence towards O) must be equal to $M_1(\mu a)\Omega^2$, since μa is the radius of its circular orbit; hence it is straightforward to show that

$$\Omega^2 = \frac{G(M_1 + M_2)}{a^3}. \tag{3.52}$$

Now Eqs. (3.7) and (3.8) hold for this system in the rotating frame, where the gravitational potential ψ is given by the sum of the potentials due to the two stars. Choosing Cartesian coordinates (x, y, z) such that the angular velocity of the frame is in the z-direction, with the stars at $(\mu a, 0, 0)$ and

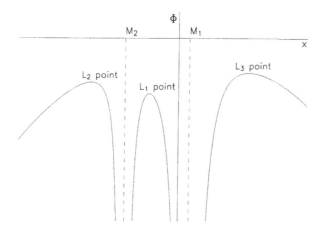

Fig. 3.2   A cut through the Roche potential of two stars with masses M_1 and M_2 along the line joining the two stars. The stars are located at the positions indicated by dashed lines. The stationary points, indicated by L_1, L_2 and L_3, are Lagrangian points.

$(-(1-\mu)a, 0, 0)$, Φ can be written from Eq. (3.8) as

$$\Phi = \frac{-GM_1}{((x-\mu a)^2 + y^2 + z^2)} - \frac{GM_2}{((x + (1-\mu)a)^2 + y^2 + z^2)}$$
$$- \frac{1}{2}\Omega^2(x^2 + y^2), \tag{3.53}$$

which is called the Roche potential. Here we have made the same approximation as in Section 3.3, namely that we can use the undistorted gravitational potential of each star: this is reasonable for centrally condensed stars in which most of the mass is concentrated near the centre.

The Roche potential (3.53) is illustrated as a function of x along the line $y = z = 0$, i.e. along the line joining the centres of two stars, in Fig. 3.2. The Lagrangian points L_1, L_2 and L_3, where $\nabla\Phi = 0$, are indicated. These are equilibrium points in the rotating frame, where the forces of attraction towards the two stars and the centrifugal force are in balance.

As in the case of a single star, the surface of each star in the binary system is a surface of constant Φ. Now provided the surface potential of each star is less that the potential Φ_{L_1} at the L_1 Lagrangian point, each star occupies a well in the Roche potential and the stars form a detached binary

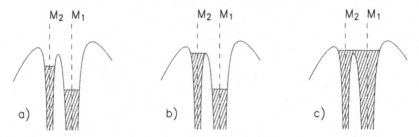

Fig. 3.3   Cuts through the Roche potential of stars along the line joining the two stars, illstrating three cases. (a) The two stars (indicated by hatching) form a detached binary system. (b) One star has filled its Roche lobe and is now losing mass to its companion star. (c) The two stars form a contact binary in which the two stellar cores occupy in a common envelope.

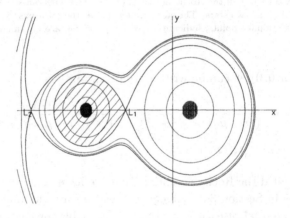

Fig. 3.4   A contour plot of the Roche potential of two stars, in a plane containing the two stars. The Roche lobe of the star on the left is indicated by the hatching.

system (Fig. 3.3a). Suppose though that M_2 expands (perhaps attempting to become a red giant) until its surface potential is equal to Φ_{L_1} (Fig. 3.3b). Any further expansion will cause matter to fall from star 2 to star 1, since it will fall to the lower potential. Algol is an example of such a binary. Finally, if the surface potentials of both stars are greater than Φ_{L_1}, then we have a "common envelope binary" or "contact binary" (Fig. 3.3c). It is also instructive to plot contours of constant Φ in the $x - y$ plane (Fig. 3.4): the shaded region, called a Roche lobe, is the maximum region the star can occupy before it starts to lose mass to its companion.

Chapter 4

Fluid Dynamical Instabilities

It is hardly possible to study astrophysical fluid dynamics without considering fluid dynamical instabilities. A fluid flow, either in reality or in a model, may be unstable, in which case perturbations may grow quickly and change the fluid configuration and its flow. Fluid dynamics involve various instabilities, which may profoundly affect the structure and evolution of astrophysical objects. A well known example is the convective instability which leads to convective cores and convective envelopes in many stars, the former affecting their nuclear evolution and the latter leading to a variety of phenomena including magnetic activity cycles in the Sun and other late-type stars.

The topic of fluid dynamical instabilities is a large one and we shall only discuss selected instabilities here. We mention a few others elsewhere in this book: in particular the Jeans instability in a self-gravitating fluid is treated in detail in Chapter 10. Excellent further reading on fluid instabilities are the book by Drazin & Reid (1981) and, particularly for rotating stars, the review by Zahn (1993).

4.1 Convective Instability

4.1.1 *The Schwarzschild criterion*

Convection plays an important role in stellar interiors and planetary atmospheres. Consider a fluid at rest with density stratification $\rho(z)$ and with gravity $g = -ge_z$ acting "downwards", so z increases upwards. The pressure distribution in z is given by hydrostatic equilibrium. Let us now consider what happens if a fluid parcel is displaced slightly upwards from height $z = z_0$ by an amount δz. We suppose that the displacement occurs sufficiently slowly that the fluid parcel remains in pressure equilibrium with its new surroundings. On the other hand we suppose that the displacement occures sufficiently quickly that no heat is exchanged between the parcel and its surroundings, so the properties of the parcel change adiabatically. The pressure and density at the original position $z = z_0$ are p_0 and ρ_0, say. At the new position $z = z_0 + \delta z$, the pressure and density of the parcel are $p_0 + \delta p$ and $\rho_0 + \delta \rho$, say. Now at $z_0 + \delta z$ the pressure of the surroundings and hence also of the fluid parcel is $p_0 + \delta z \mathrm{d}p/\mathrm{d}z$ to first order in δz; so $\delta p = \delta z \mathrm{d}p/\mathrm{d}z$. The density of the surroundings is $\rho_0 + \delta z \mathrm{d}\rho/\mathrm{d}z$. But by the adiabatic assumption, the parcel's pressure and density perturbations are related by

$$\frac{p_0 + \delta p}{p_0} = \left(\frac{\rho_0 + \delta \rho}{\rho_0} \right)^{\gamma}. \tag{4.1}$$

Hence, linearizing in perturbation quantities, the density perturbation of the parcel is

$$\delta \rho = \frac{\rho_0}{\gamma p_0} \delta p = \frac{\rho_0}{\gamma p_0} \delta z \frac{\mathrm{d}p}{\mathrm{d}z}. \tag{4.2}$$

The parcel finds itself heavier than its surroundings and hence sinks back towards its original location if

$$\rho_0 + \delta \rho > \rho_0 + \delta z \frac{\mathrm{d}\rho}{\mathrm{d}z}, \tag{4.3}$$

i.e., if

$$\frac{\rho_0}{\gamma p_0} \delta z \frac{\mathrm{d}p}{\mathrm{d}z} > \delta z \frac{\mathrm{d}\rho}{\mathrm{d}z},$$

i.e.,

$$\frac{1}{\gamma p} \frac{\mathrm{d}p}{\mathrm{d}z} > \frac{1}{\rho} \frac{\mathrm{d}\rho}{\mathrm{d}z}, \quad i.e., \quad \frac{1}{\gamma} < \frac{\mathrm{d}\ln \rho}{\mathrm{d}\ln p}$$

(since $\mathrm{d}p/\mathrm{d}z < 0$). Conversely, if

$$\frac{1}{\gamma} > \frac{\mathrm{d}\ln \rho}{\mathrm{d}\ln p} \tag{4.4}$$

then the parcel finds itself lighter than the surrounding fluid and hence continues rising. In the latter case, the original stratification is unstable and fluid parcels will move around, i.e. the stratification is convectively unstable. Note that the density can increase with depth (equivalently, increase with pressure) and still be unstable: it has to increase sufficiently rapidly with depth to be stable to convection. Exactly the same criterion would result from considering the parcel moving downwards instead, though one must then remember to reverse the inequality when dividing by the (negative) δz.

The energy source for the convective instability is the potential energy of the original unstable stratification.

If the stratification is stable to the above criterion then the acceleration of the parcel is given by

$$\rho \frac{d^2 \delta z}{dt^2} = \left(\rho + \delta z \frac{\partial \rho}{\partial z} \right) g - (\rho + \delta \rho) g \tag{4.5}$$

(buoyancy force minus weight), where we now consider δz to be a function of time; and hence

$$\frac{d^2 \delta z}{dt^2} + N^2 \delta z = 0 \tag{4.6}$$

where

$$N^2 = g \left(\frac{1}{\gamma} \frac{d \ln p}{dz} - \frac{d \ln \rho}{dz} \right) = \frac{\rho g^2}{p} \left(\frac{d \ln \rho}{d \ln p} - \frac{1}{\gamma} \right). \tag{4.7}$$

(Recall that z increases upwards: if instead z had been defined to increase downwards, the signs of the two derivatives in z would have ben reversed.) Equation (4.6) describes simple harmonic motion with frequency N. Thus the parcel can oscillate in the vertical direction about its equilibrium position with frequency N, which is known as the *Brunt-Väisälä frequency* or *buoyancy frequency*. This is the mechanism for internal gravity waves in a stably stratified fluid (e.g. Section 12.4).

The instability criterion (4.4) is thus that

$$N^2 < 0 \tag{4.8}$$

(which would imply oscillations with imaginary frequency, one root leading to exponential growth of the displacement from equilibrium).

If the chemical composition is uniform, then

$$\ln p = \ln \rho + \ln T + \text{constant}$$

for a perfect gas (see Section 1.7) and so the instability criterion (4.4) becomes

$$\frac{d \ln T}{d \ln p} > 1 - \frac{1}{\gamma} \tag{4.9}$$

which is often written

$$\nabla > \nabla_{\text{ad}} \tag{4.10}$$

for the fluid to be unstable, where

$$\nabla \equiv \frac{d \ln T}{d \ln p} \quad \text{and} \quad \nabla_{\text{ad}} \equiv \left(\frac{d \ln T}{d \ln p} \right)_S = 1 - \frac{1}{\gamma} . \tag{4.11}$$

Equation (4.9) is the Schwarzschild criterion for convective instability.

More generally, if there is a vertical gradient of chemical composition, characterized by

$$\nabla_\mu \equiv \frac{d \ln \mu}{d \ln p} , \tag{4.12}$$

then

$$N^2 = \frac{g\delta}{H_p} (\nabla_{\text{ad}} - \nabla + \nabla_\mu) \tag{4.13}$$

where H_p is the pressure scale height (2.70) and

$$\delta \equiv \left(\frac{\partial \ln \rho}{\partial \ln T} \right)_p \tag{4.14}$$

($\delta = 1$ for a perfect gas). Then the criterion for instability is

$$\nabla > \nabla_{\text{ad}} + \nabla_\mu ; \tag{4.15}$$

this is the Ledoux criterion for convective instability (see Section 4.1.2).

The Schwarzschild instability criterion for a region of uniform composition can also be expressed in terms of the gradient of specific entropy S, noting that

$$\frac{\delta}{c_p} dS = \frac{1}{\gamma} \frac{dp}{p} - \frac{d\rho}{\rho} \tag{4.16}$$

where c_p is the specific heat at constant pressure. Then from (4.7),

$$N^2 = \frac{g\delta}{c_p} \frac{dS}{dz} ; \tag{4.17}$$

the layer is stable to convection if the specific entropy increases upwards.

If $\nabla > \nabla_{\text{ad}}$ the temperature gradient is said to be *superadiabatic*; if $\nabla < \nabla_{\text{ad}}$ it is *subadiabatic*. Convective motions transport heat: commonly in stellar interior regions that are convectively unstable, convection is very efficient in that it requires only a very small superadiabatic gradient $\nabla - \nabla_{\text{ad}}$

to transport the star's entire heat flux. In that case $\nabla \simeq \nabla_{\text{ad}}$ and $N^2 \simeq 0$ there; also S is nearly constant.

Most stellar-structure modelling uses a simple phenomenological description called *mixing-length theory* to calculate the heat transport by convection and hence the stratification required in convectively unstable regions in order to produce the necessary convective heat flux (e.g. Kippenhahn & Weigart 1990). The idea is that blobs of convected fluid travel a distance ℓ from their position of equilibrium and then disrupt and disperse into the new surroundings: ℓ is the mixing length. The mixing length has to be prescribed. In stellar convective envelopes it is commonly presumed to be a fixed constant times the local pressure scale height: the value of the fixed constant (the *mixing-length parameter*) can be adjusted so as produce e.g. a solar model of the correct radius. Of course, since there is no real theory involved, we do not know that the mixing-length parameter should be the same for different stars. Other prescriptions of the mixing length are possible. In order to use this mixing-length theory in modelling stellar structure, it is necessary to calculate the speed at which blobs move and the amount of heat they transport. It should borne in mind that this is all fairly crude and that the adjustment of the mixing-length parameter takes up the slack left by inexactitude in the argument. Thus the speed v of the blobs can be calculated from the equation of motion (4.6) (where N^2 is negative for an unstable region) assuming at its simplest that the blob starts from rest and travels a distance ℓ at constant acceleration. The convective heat transport effected by the motion of the blobs is $\rho v c_p \Delta T$ where c_p is the specific heat capacity and ΔT is the temperature excess of a blob over its surroundings. Since $\Delta p = 0$ (the blob is in pressure equilibrium with its surroundings), the temperature excess can be written in terms of the superadiabatic gradient divided by the pressure scale height, multiplied by the distance ℓ travelled by the blob. Since it turns out that v is proportional to the square root of the superadiabatic gradient, the convective heat flux is proportional to the superadiabatic gradient raised to the power $3/2$. As already stated, in deep convective envelopes of stars it turns out that the specific heat capacity is large enough that a tiny superadiabatic temperature gradient is sufficient to give the necessary convective heat flux, so that $N^2 \approx 0$ there.

4.1.2 *Effects of dissipation*

Convective instability is a *dynamical process*: it does not require a dissipative process, and can therefore be treated in the adiabatic, inviscid approximation. As established by Rayleigh (1880), dissipation modifies the instability criterion only slightly:

$$N^2 < -\frac{C}{t_v t_d} \tag{4.18}$$

for instability. Here C is a positive constant of order unity, and t_v and t_d are the dissipation timescales associated with viscosity and heat diffusion, respectively (see Zahn 1993). Constant C depends on the geometry of the fluid region and on boundary conditions. Inside stars, t_d is shorter than t_v since the thermal diffusivity κ is generally much larger than the viscosity ν; the Prandtl number $\text{Pr} \equiv \nu/\kappa$ is estimated to be of order 10^{-9} to 10^{-6} in the Sun: see Lignières (1999). In the convectively stable regions, diffusion tends to damp oscillations. When t_d ($\sim l^2/\kappa$) becomes comparable with N^{-1}, where l is a characteristic parcel size to be damped, the entropy stratification is no longer effective in stabilizing the layer, though composition gradients if present will still provide a restoring force.

Interesting competing effects ("double-diffusive instability") occur in a layer which is dynamically stable, i.e. $N^2 > 0$ in (4.13) but either the composition stratification or thermal stratification on their own would be unstable. The archetypal laboratory example is water heated either from above or below and with a gradient in salinity. Heat diffuses much faster than the salt concentration. If it is the salinity gradient that is destabilizing (salt water on top of fresh water, but with the top of the layer hotter than the bottom), then small-scale perturbations for which $t_d \sim N^{-1}$ can grow, producing so-called salt fingers. Eventually the stratification becomes a layered convection with both temperature and salinity varying stepwise in depth – see Zahn (1993). The opposite case is that of salty water beneath fresh water, with an otherwise unstable temperature gradient which can be produced by heating the water from below. One can envisage a displaced fluid parcel oscillating but cooling down (due to the shorter timescale t_d when it is above its equilibrium position and heating up when below the equilibrium position. This causes the velocity at which the parcel passes the equilibrium level to increase and the amplitude of the oscillation to grow. This is *overstability*: this particular example is called thermohaline convection.

In stellar cores, helium may play the analogous role to the salt in the last example. The stellar stratification may be thermally unstable so the Schwarzschild criterion (4.10) for instability is satisfied, but stable overall so that the Ledoux criterion (4.15) for instability is not satisfied. The resulting motion is sometimes referred to as *semi-convection* in stellar astrophysics.

As pointed out by Zahn (1993), angular momentum may also play the analogous role to the salt, since it too diffuses more slowly than heat.

4.1.3 *Modelling convection: the Boussinesq approximation*

Before leaving the topic of convection, we mention a commonly adopted approximation in modelling convectively unstable regions. This is the *Boussinesq approximation*. Let p' be the fluctuation of pressure about its horizontal mean, and similarly for other theromodynamic quantities. The velocity u is also considered a fluctuation, about a reference state in which the velocity is zero. In the Boussinesq approximation the density fluctuations are retained only in the buoyancy term in the equation of motion. Specifically, then, the density fluctuations are ignored in the continuity equation. The latter is called the anelastic approximation (Gough 1969) and filters out high-frequency phenomena such as sound waves that may be considered unimportant for transport properties of the flow. The set of equations for fluctuations in the Boussinesq approximation is

$$\frac{\partial u}{\partial t} + u \cdot \nabla u = -\frac{1}{\rho}\nabla p' - \frac{\rho'}{\rho}g + \nu\nabla^2 u \qquad (4.19)$$

$$\nabla \cdot u = 0 \qquad (4.20)$$

$$\frac{\partial T'}{\partial t} + u \cdot \nabla T' - \beta e_z \cdot u = \text{radiative exchange term} \qquad (4.21)$$

(e.g. Gough 1977) where $\beta = (T/H_p)(\nabla - \nabla_{\text{ad}})$ is the so-called superadiabatic lapse rate. Moreover, in the Boussinesq approximation the ρ'/ρ in the first equation gets replaced by $\delta T'/T$, where $\delta = -(\partial \ln \rho/\partial \ln T)_p$, i.e. pressure fluctuations are neglected compared to temperature fluctuations.

The Boussinesq approximation can be justified in laboratory convection, where the scale heights of pressure and density are long compared to the depth of the fluid layer under study, and also in some geophysical applications. Its application to stellar convection is not really justifiable, but it does yield a more tractable problem and may yield some insight into the full problem. See Spiegel (1971) for a discussion of stellar convection and the Boussinesq approximation.

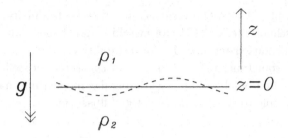

Fig. 4.1 The set-up for the Rayleigh-Taylor instability: one fluid of (uniform) density ρ_1 overlying another of density ρ_2. The gravitational acceleration is g and z is the vertical coordinate (height). The dashed curve represents the perturbed interface between the two fluids.

4.2 The Rayleigh-Taylor Instability

Several instabilities can occur at interfaces between fluids. Consider two fluids of uniform (but different) density with a plane interface between them, with a uniform gravitational field perpendicular to the interface (Fig. 4.1). If the denser fluid is on top, (so $\rho_1 > \rho_2$ in the notation defined in the figure), the Rayleigh-Taylor instability develops. This may seem an unlikely configuration to occur in nature, but g can equally be an *effective* gravitational acceleration, e.g. at an accelerating shock front and this situation can be found in supernovae, for example. The Rayleigh-Taylor instability commonly occurs at the same time as the Kelvin-Helmholtz instability, which arises from velocity shear between the two layers: the Kelvin-Helmholtz instability is discussed in Section 4.4.

To understand the development of the Rayleigh-Taylor instability, we consider what happens if there is a small perturbation to the interface. If the perturbation grows then the configuration is unstable. Since the background configuration is translationally invariant horizontally (Fig. 4.1), we may without loss of generality consider an individual Fourier component, in the x-direction say, so with x-dependence e^{ikx}. Likewise the time-independence of the background means that we seek a temporal variation of the form $e^{-i\omega t}$ say. We suppose that any velocities, pressure variations, etc. arise only from the perturbation from the interface: hence all perturbation variables will be proportional to

$$\exp(ikx - i\omega t) \,. \tag{4.22}$$

Without loss of generality we take k to be positive.

We shall analyse a more general situation in Section 4.4, so rather than repeat the mathematical analysis we simply here quote the result: from Eq. (4.38) with U_1 and U_2 set to zero, we find that for irrotational, incompressible perturbations the temporal frequency ω of the perturbation is related to its horizontal wavenumber k by

$$\omega^2 = \left(\frac{\rho_2 - \rho_1}{\rho_2 + \rho_1} \right) gk . \tag{4.23}$$

Thus if $\rho_1 > \rho_2$ (heavier fluid on top) $\omega^2 < 0$ and so ω is imaginary: one of the two roots corresponds to exponentially growing solutions (4.22), and so the configuration is unstable. This is the *Rayleigh-Taylor instability*.

We note in passing that if we put $\rho_1 = 0$ then we recover $\omega^2 = gk$, which is the dispersion relation for surface gravity waves that was derived in Section 2.6. Expression (4.23) generalizes that dispersion relation to include a uniform-density upper layer.

As with the convective instability (Section 4.1), the energy source for the Rayleigh-Taylor instability is the potential energy stored in the initial configuration.

4.3 Rotational Instability

Rotation introduces a whole new range of possible instabilities. Some are associated with shear, and shall be considered in the next section. For now we consider only a simple scenario, which we might envisage in the interior of a star, say, in which the rotation rate Ω varies only with the distance ϖ from the axis (so ϖ is the radial coordinate in a cylindrical polar coordinate set). We assume that effects of viscosity are negligible. We also for simplicity neglect gravity, so in the equilibrium configuration pressure and centrifugal forces balance:

$$\frac{1}{\rho} \frac{\mathrm{d}p}{\mathrm{d}\varpi} + \varpi \Omega^2 = 0 . \tag{4.24}$$

Consider now a parcel of fluid undergoing a small radial displacement from ϖ to $\varpi + \delta\varpi$. Since viscosity is negligible, the parcel conserves its specific angular momentum $h \equiv \varpi^2 \Omega$. On the other hand, the pressure force the parcel feels in its new surroundings is determined by the angular velocity there, according to Eq. (4.24). Hence the imbalance of force per unit mass in the radial direction felt by the parcel after its displacement is

$$(\varpi + \delta\varpi)\left\{\frac{\varpi^2\Omega(\varpi)}{(\varpi + \delta\varpi)^2}\right\} - (\varpi + \delta\varpi)\left\{\Omega(\varpi + \delta\varpi)\right\}$$
$$\equiv -\frac{1}{\varpi^3}\frac{\mathrm{d}}{\mathrm{d}\varpi}\left(\varpi^4\Omega^2\right)\delta\varpi \equiv -N_\Omega^2\,\delta\varpi \qquad (4.25)$$

to $O(\delta\varpi)$. Thus the equation of motion of the parcel is

$$\frac{\mathrm{d}^2\delta\varpi}{\mathrm{d}t^2} + N_\Omega^2\delta\varpi = 0 \qquad (4.26)$$

which gives a stable, oscillatory motion if N_Ω^2 is positive but an instability if N_Ω^2 is negative. As in the convective and Rayleigh-Taylor instabilities, this instability is a dynamical instability. Thus for stability the rotation profile must satisfy

$$\frac{1}{\varpi^3}\frac{\mathrm{d}}{\mathrm{d}\varpi}\left(\varpi^4\Omega^2\right) > 0\,. \qquad (4.27)$$

Rather strong differential rotations are required to trigger this instability and in practice other instabilities would usually set in sooner (Zahn 1993). Such a shear instability is discussed next.

4.4 Shear and the Kelvin-Helmholtz Instability

Shear instabilities, which can occur when a fluid's velocity is not uniform, are very common. Their main feature is that they extract vorticity from a laminar (i.e. non-turbulent) flow and the vortices can then grow and interact with one another. An excellent reference is the book by Drazin & Reid (1981); Zahn (1993) also provides a good reference.

Here we restrict ourselves to an analysis of the Kelvin-Helmholtz instability, and some general comments about the onset of shear instabilities and about turbulence.

4.4.1 *The Kelvin-Helmholtz instability*

The Kelvin-Helmholtz instability can occur when one fluid flows over another. Consider a two-layer fluid with a planar interface ($z = 0$) in a uniform gravitational field $g = -g e_z$. Each fluid has its own uniform density and uniform, steady, horizontal velocity (in the x-direction say). The configuration is illustrated in Fig. 4.2.

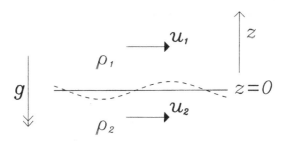

Fig. 4.2 The set-up for the Kelvin-Helmholtz instability: one fluid layer flowing over another. The upper fluid layer has density ρ_1 and horizontal speed u_1, the lower layer has density ρ_2 and horizontal speed u_2. For other details, see the caption to Fig. 4.1.

We have already seen that in the case of no horizontal velocity such a configuration is subject to the Rayleigh-Taylor instability if the fluid in top is more dense than the fluid below ($\rho_1 > \rho_2$). We shall now see that if there is shear between the two layers (i.e. $U_1 \neq U_2$) then this further destabilizes the configuration, which may then be unstable even if the lower layer is the denser one. Put another way, a sufficiently stable density stratification is necessary to overcome the shear instability.

As in Section 4.2, we consider a perturbation $z = \zeta(x)$ of the interface of the form

$$\zeta = A \exp\left(ikx - i\omega t\right) . \tag{4.28}$$

We shall consider incompressible, irrotational perturbations in each layer, so that the small perturbations $\boldsymbol{u}'$ to the background velocity are expressible in terms of a scalar potential ϕ:

$$\boldsymbol{u}' = \nabla\phi , \qquad \nabla^2\phi = 0 . \tag{4.29}$$

We shall use subscripts 1 and 2 to denote quantities in the upper and lower layers respectively; so, for example, the total velocity $\boldsymbol{u}$ is $U_1\boldsymbol{e}_x + \nabla\phi_1$ in the upper layer and $U_2\boldsymbol{e}_x + \nabla\phi_2$ in the lower one.

All perturbations are deriven by the interface distortion. Thus all perturbed quantities have the same dependence on x and t as in Eq. (4.28); and they all tend to zero far from the interface. Without loss of generality we take k to be positive. Then since the velocity potentials are solutions of

Laplace's equation (4.29), one can immediately say that

$$\begin{aligned}
\phi_1 &= C_1 \exp(-i\omega t + ikx - kz)\,, \\
\phi_2 &= C_2 \exp(-i\omega t + ikx + kz)
\end{aligned} \tag{4.30}$$

for some constants C_1 and C_2. The different signs in the z-dependence ensures that $\phi \to 0$ as $z \to \pm\infty$.

A kinematic condition is that, on either side of the interface, any fluid particle at the interface surface must remain on the surface which means that the vertical component of the velocity must match the material derivative of the interface displacement $\zeta(x,t)$:

$$\begin{aligned}
\frac{\partial \phi_1}{\partial z} &= \frac{\partial \zeta}{\partial t} + U_1 \frac{\partial \zeta}{\partial x} \\[4pt]
\frac{\partial \phi_2}{\partial z} &= \frac{\partial \zeta}{\partial t} + U_2 \frac{\partial \zeta}{\partial x}
\end{aligned} \tag{4.31}$$

at the interface. Correct to first order, since terms in these equations are already first-order small quantities, we may evaluate Eqs. (4.31) at the unperturbed surface location $z = 0$. Thus using Eqs. (4.28) and (4.30) we deduce that

$$-k\,C_1 = -i\omega A + ikU_1 A\,, \quad k\,C_2 = -i\omega A + ikU_2 A\,. \tag{4.32}$$

Another condition is that the normal stress across the interface must be continuous, which here means that the pressure p must be continuous. Now the momentum equation can be written

$$\nabla\left(\frac{\partial \phi}{\partial t}\right) + \nabla\left(\frac{1}{2}\boldsymbol{u}^2\right) = -\frac{1}{\rho}\nabla p - g\boldsymbol{e}_z \tag{4.33}$$

which can be integrated to give (to linear order)

$$\frac{\partial \phi}{\partial t} + U\frac{\partial \phi}{\partial x} = -\frac{p}{\rho} - gz + F(t)\,, \tag{4.34}$$

where $F(t)$ is a "constant" of integration. Now since the only time-dependent quantities are the perturbations, and these tend to zero far from the interface, we can deduce that $F(t)$ is identically zero. Thus continuity of pressure at the interface implies, using Eq. (4.34), that

$$-\rho_1\left(\frac{\partial \phi_1}{\partial t} + U_1\frac{\partial \phi_1}{\partial x} + g\zeta\right) = -\rho_2\left(\frac{\partial \phi_2}{\partial t} + U_2\frac{\partial \phi_2}{\partial x} + g\zeta\right). \tag{4.35}$$

Again it is now consistent to evaluate this at $z = 0$; hence

$$\rho_1 \left(-i\omega C_1 + ikU_1C_1 + gA\right) = \rho_2 \left(-i\omega C_2 + ikU_2C_2 + gA\right) . \qquad (4.36)$$

Substituting for C_1 and C_2 from Eqs. (4.32) gives an equation that is homogeneous in A: for non-trivial ($A \neq 0$) solutions we must have that

$$\rho_1 \left(\omega - kU_1\right)^2 + \rho_1 gk = -\rho_2 \left(\omega - kU_2\right)^2 + \rho_2 gk , \qquad (4.37)$$

i.e.,

$$\left(\omega - k\bar{U}\right)^2 = \frac{(\rho_2 - \rho_1)gk}{(\rho_1 + \rho_2)} - \frac{\rho_1\rho_2(U_1 - U_2)^2k^2}{(\rho_1 + \rho_2)^2} \qquad (4.38)$$

where $\bar{U} = (\rho_1 U_1 + \rho_2 U_2)/(\rho_1 + \rho_2)$ is a density-weighted average speed.

The configuration is unstable if the right-hand side of Eq. (4.38) is negative, since then ω will have a non-zero imaginary part and one of the two solutions will correspond to exponential growth with time. If $U_1 = U_2$ we obtain the criterion (4.23) for the Rayleigh-Taylor instability. If $U_1 \neq U_2$, we see that the configuration is stable only if the first (stratification) term on the right-hand side of (4.38) is larger than the second, shear, term which always acts in the sense to destabilize the system. Disturbances of sufficiently small wavelength (high k) are always unstable if $U_1 \neq U_2$, though this is not true in a real layer of finite thickness, see Drazin & Reid (1981). Note that the $\bar{U}$ term on the left of Eq. (4.38) is merely equivalent to a Galilean transformation in the x-direction with speed $\bar{U}$.

If the shear causes the configuration to be unstable by making the right of (4.38) negative, this is called the *Kelvin-Helmholtz instability*. Note that we have only proved a sufficient condition for stability since we have considered only a subset of possible disturbances, namely irrotational ones.

4.4.2 *Critical Richardson and Reynolds numbers*

We have seen in Section 4.4.1 how a stable stratification hinders the onset of shear instablity. In a more general configuration, where stratification and velocity vary with height z, the stabilizing effect is measured by the Richardson number

$$\mathrm{Ri} \equiv \frac{N^2}{(\mathrm{d}U/\mathrm{d}z)^2} . \qquad (4.39)$$

In the absence of dissipation, a sufficient condition for instability for a variety of velocity and density profiles is that Ri is smaller than a critical value which is 1/4. If heat can dissipate, however, this will weaken the buoyancy force (see the discussion of double-diffusion, Section 4.1.2) and make the layer less stable; thus the critical Richardson number is increased. See Zahn (1993) for a fuller discussion of the issue.

Another important quantitity determining the onset of turbulence in a viscous flow is the critical Reynolds number. The Reynolds number is defined by

$$\text{Re} \equiv LU/\nu \qquad (4.40)$$

where L and U are characteristic lengthscale and speed of the flow respectively, and ν is the kinematic viscosity. The Reynolds number measures the relative importance of inertial terms and the viscous term in the momentum equation. In the absence of other forces (e.g. buoyancy), a laminar flow becomes unstable when the Reynolds number exceeds some critical number Re_c. Generally Re_c is of the order of 1000, but it depends on boundary conditions of the flow and on the particular velocity profile. See Drazin & Reid (1981).

4.4.3 *Turbulence and the Kolmogorov spectrum*

Instabilities such as those discussed above can rapidly lead in high Reynolds-number flow to turbulence, in which neighbouring parcels of fluid at some instant rapidly follow very different and practically unpredictable trajectories. (A flow that is not turbulent is called laminar.) The onset of turbulence may be pictured as the flow developing smaller and smaller scales of motion until at sufficiently small scales molecular viscosity sets in. Turbulence is a very challenging problem. We shall not consider further the stages by which turbulence develops, but focus instead on fully developed turbulence. Even there the challenges are formidable and we restrict our attention to the case where the statistical properties of the turbulence are homogeneous, isotropic and steady. The basic ideas were conceived by Kolmogorov (1941) and a classic text in which the theory is developed is Batchelor (1953).

We envisage that the turbulent flow is in a statistically steady state. Energy enters the flow at a rate ϵ in motions on some lengthscale l_0. The energy per unit mass on this lengthscale is $\frac{1}{2}u_0^2$. This energy then cascades to smaller and smaller scales l until eventually at some scale l_ν it is dissipated as heat by the fluid's molecular viscosity ν.

It seems reasonable to suppose that dissipative lengthscale l_ν does not depend on l_0 but only on ϵ and ν. Now the dimensions of ϵ are $[\epsilon] = L^2 T^{-3}$, where L and T denote dimensions of length and time; and the dimensions of ν are $[\nu] = L^2 T^{-1}$. Therefore on dimensional grounds we deduce that

$$l_\nu \sim \left(\nu^3/\epsilon\right)^{1/4} . \tag{4.41}$$

The kinetic energy per unit mass on scale l_0 presumably depends only on l_0 and ϵ; thus again by dimensional arguments

$$u_0^2 \sim (\epsilon l_0)^{2/3} . \tag{4.42}$$

Eliminating ϵ between these two expressions yields

$$l_\nu \sim (\mathrm{Re})^{-3/4} l_0 \tag{4.43}$$

with Reynolds number $\mathrm{Re} \equiv u_0 l_0/\nu$.

We also introduce wavenumbers k so

$$k = l^{-1}, \quad k_\nu = l_\nu^{-1}, \quad k_0 = l_0^{-1} . \tag{4.44}$$

The interval $k_0 < k < k_\nu$ in wavenumber space is called the *inertial range*.

The *kinetic energy spectrum* $E(k,t)$ is defined such that the average kinetic per unit mass is

$$\frac{1}{2}\langle u^2 \rangle = \int_0^\infty E \, \mathrm{d}k . \tag{4.45}$$

Because of viscous dissipation, there is an upper cut-off to the integral at $k = k_\nu$. On dimensional grounds once more, since $[E] = L^3 T^{-2}$, we deduce that in the inertial range

$$E = C \epsilon^{2/3} k^{-5/3} \tag{4.46}$$

where C is a constant, assuming that E is independent l_0 and l_ν there. This is the Kolmogorov scaling, that in the inertial range the energy density in homogeneous turbulence scales as $k^{-5/3}$.

Chapter 5

Magnetohydrodynamics

In this chapter we consider some of the new features introduced in a fluid in the presence of a magnetic field. We only consider applications where the continuum fluid approximation is still applicable.

Magnetohydrodynamics is the study of the motion of an electrically conducting fluid in the presence of a magnetic field. Ionized fluids and magnetic fields are common in astrophysics. The presence of a magnetic field is significant to the fluid equilibrium and dynamics if the *plasma beta* $\beta = p/(B^2/2\mu_0) \lesssim 1$, where p is the fluid (commonly gas) pressure, B is the strength of the magnetic field, and μ_0 is the magnetic permeability: β is the ratio of fluid pressure to magnetic pressure $(B^2/2\mu_0)$. Astrophysical areas of interest include: planetary, stellar and galactic dynamos; the solar corona; planetary magnetospheres; and solar and stellar winds.

5.1 Maxwell's Equations and the MHD Approximation

In the presence of a magnetic field $\boldsymbol{B}$, a conducting fluid will be subject to an additional force per unit volume, the Lorentz force $\boldsymbol{j} \times \boldsymbol{B}$, where $\boldsymbol{j}$ is the current density. Hence the momentum equation for such a fluid, assuming the only other forces are the pressure gradient and gravity, is

$$\rho \frac{D\boldsymbol{u}}{Dt} = -\nabla p + \rho \boldsymbol{g} + \boldsymbol{j} \times \boldsymbol{B} \, . \tag{5.1}$$

The equations of fluid dynamics must be supplemented by Maxwell's equations. These are

$$\nabla \cdot \boldsymbol{E} = \frac{\rho_e}{\epsilon_0} \tag{5.2}$$

$$\nabla \cdot \boldsymbol{B} = 0 \tag{5.3}$$

$$\nabla \times \boldsymbol{E} = -\frac{\partial \boldsymbol{B}}{\partial t} \tag{5.4}$$

$$\nabla \times \boldsymbol{B} = \mu_0 \boldsymbol{j} + \frac{1}{c^2} \frac{\partial \boldsymbol{E}}{\partial t} \tag{5.5}$$

where $\boldsymbol{E}$ is the electric field, ρ is the electric charge density , and ϵ_0 is the permittivity; and, in this subsection only, c is the speed of light.

The second of Maxwell's equations, $\nabla \cdot \boldsymbol{B} = 0$, implies that $\boldsymbol{B}$ can be written in terms of a vector potential $\boldsymbol{A}$, such that $\boldsymbol{B} = \nabla \times \boldsymbol{A}$.

We shall work entirely in the magnetohydrodynamic (MHD) approximation, so assuming that (i) the electron and ion fluids are locked together; (ii) relativistic effects are negligible; (iii) $\boldsymbol{j} \simeq \boldsymbol{j}' = \sigma \boldsymbol{E}' \simeq \sigma(\boldsymbol{E} + \boldsymbol{u} \times \boldsymbol{B})$ where primes (here only) denote quantities in a frame co-moving with the fluid. Also σ is the electrical conductivity. In this approximation, the displacement current – the last term in the fourth Maxwell equation – is negligible and so this equation yields

$$\boldsymbol{j} = \frac{1}{\mu_0} \nabla \times \boldsymbol{B} \tag{5.6}$$

and so the Lorentz force $\boldsymbol{j} \times \boldsymbol{B}$ can be expressed as $\mu_0^{-1}(\nabla \times \boldsymbol{B}) \times \boldsymbol{B}$. Also the third Maxwell equation yields

$$\frac{\partial \boldsymbol{B}}{\partial t} = \nabla \times (\boldsymbol{u} \times \boldsymbol{B}) + \eta \nabla^2 \boldsymbol{B} \tag{5.7}$$

where $\eta = 1/\mu_0 \sigma$ is the magnetic diffusivity (assumed uniform). Equation (5.7) is the magnetic induction equation: note the parallel with the evolution equation for vorticity:

$$\frac{\partial \boldsymbol{\omega}}{\partial t} = \nabla \times (\boldsymbol{u} \times \boldsymbol{\omega}) + \nu \nabla^2 \boldsymbol{\omega} . \tag{5.8}$$

In many astrophysical applications the high conductivity limit applies, so $\eta \simeq 0$ and the diffusion term in Eq. (5.7) can be neglected. Then the induction equation becomes

$$\frac{\partial \boldsymbol{B}}{\partial t} = \nabla \times (\boldsymbol{u} \times \boldsymbol{B}) . \tag{5.9}$$

In terms of vector potential $\boldsymbol{A}$ this yields

$$\frac{\partial \boldsymbol{A}}{\partial t} = \boldsymbol{u} \times (\nabla \times \boldsymbol{A}) , \quad i.e., \quad \frac{D A_i}{Dt} = u_j \frac{\partial A_j}{x_i} . \tag{5.10}$$

Equation (5.9) and the continuity equation can be combined to give

$$\frac{D}{Dt} \left(\frac{\boldsymbol{B}}{\rho} \right) = \left(\frac{\boldsymbol{B}}{\rho} \right) \cdot \nabla \boldsymbol{u} . \tag{5.11}$$

This is the same as the evolution equation (1.50) for a material line element $\boldsymbol{dl}$:

$$\frac{D \boldsymbol{dl}}{Dt} = \boldsymbol{dl} \cdot \nabla \boldsymbol{u} . \tag{5.12}$$

Thus magnetic field lines move as if frozen in the fluid.

Another way to view "flux freezing" is that the flux $\Phi \equiv \int_S \boldsymbol{B} \cdot \boldsymbol{dl}$ through a material surface S spanning closed material curve C is a constant. For, using Stokes's theorem twice,

$$\frac{d\Phi}{dt} = \frac{d}{dt} \int_C \boldsymbol{A} \cdot \boldsymbol{dl} = \int_C \frac{D\boldsymbol{A}}{Dt} \cdot \boldsymbol{dl} + \int_C \boldsymbol{A} \cdot \frac{D\boldsymbol{dl}}{Dt} = \int_C \nabla (\boldsymbol{u} \cdot \boldsymbol{A}) \cdot \boldsymbol{dl} = 0$$

from equations (5.10) and (5.12).

The Lorentz force term may be rewritten

$$\frac{1}{\mu_0} (\nabla \times \boldsymbol{B}) \times \boldsymbol{B} = \frac{1}{\mu_0} \boldsymbol{B} \cdot \nabla \boldsymbol{B} - \nabla \left(B^2 / 2\mu_0 \right) , \tag{5.13}$$

assuming constant μ_0. The first term on the right-hand side represents a magnetic tension (by analogy with a stretched string) and is zero if $\boldsymbol{B}$ does not change in the direction of $\boldsymbol{B}$. The second term looks like a pressure term and can be thought of as arising from an isotropic magnetic pressure $p_{\text{mag}} = B^2 / 2\mu_0$.

The presence of a magnetic field modifies the hydrostatic equilibrium. In the absence of fluid motion, equilibrium requires from Eq. (5.1) that

$$\nabla \left(p + B^2 / 2\mu_0 \right) = \rho \boldsymbol{g} + \frac{1}{\mu_0} \boldsymbol{B} \cdot \nabla \boldsymbol{B} . \tag{5.14}$$

This equation forms the basis for modelling e.g. quiescent magnetic prominences in the solar atmosphere (Section 5.3.1).

5.2   MHD Waves

The presence of a magnetic field increases the richness of small-amplitude
wave motions that can be supported by the medium. For simplicity we consider here only the case of a unidirectional equilibrium field in the absence
of gravity.

An important quantity that emerges is the Alfvén speed v_A, given by

$$v_A^2 = \frac{B^2}{\mu_0 \rho} .$$

Note that the plasma β, introduced earlier, and the Alfvén speed are related
by $\beta^{-1} = \frac{1}{2}\gamma_1 v_A^2/c^2$. Wave propagation involves both the sound speed and
the Alfvén speed. Also important are the combinations

$$c_f^2 = c^2 + v_A^2 , \quad c_T^{-2} = c^{-2} + v_A^{-2}$$

where c_f is called the fast (magnetoacoustic) speed and c_T the slow speed.
Note that c_f is greater than both c and v_A, while c_T is smaller than c and
v_A.

The linearized perturbed equations are

$$\rho \frac{\partial \boldsymbol{u}}{\partial t} = -\nabla p' + \frac{1}{\mu_0}\left(\nabla \times \boldsymbol{B}'\right) \times \boldsymbol{B}$$

$$\frac{\partial \rho'}{\partial t} + \nabla \cdot (\rho \boldsymbol{u}) = 0$$

$$\frac{\partial \boldsymbol{B}'}{\partial t} = \nabla \times (\boldsymbol{u} \times \boldsymbol{B}) \qquad (5.15)$$

$$\nabla \cdot \boldsymbol{B}' = 0$$

about an equilibrium satisfying $\nabla(p + B^2/2\mu_0) = 0$. Now as usual primes
denote perturbation quantities ($\boldsymbol{u}$ is also small). Also we have the adiabatic
energy equation

$$\frac{\partial p'}{\partial t} + \boldsymbol{u} \cdot \nabla p = c^2 \left(\frac{\partial \rho'}{\partial t} + \boldsymbol{u} \cdot \nabla \rho\right) \qquad (5.16)$$

where once more $c^2 = \gamma_1 p/\rho$ now represents the square of the adiabatic
sound speed. Without loss of generality we take $\boldsymbol{B}$ to be in the z-direction.

The simplest case to consider is a uniform magnetic field and background
medium. Then we may seek solutions $\propto \exp(ik_x x + ik_y y + ik_z z - i\omega t)$.
We write $k^2 = k_x^2 + k_y^2 + k_z^2$; without loss of generality $k_y = 0$. Then
$\partial/\partial t \to -i\omega\times$, $\partial/\partial x \to ik_x\times$, etc., and so Eqs. (5.15) become algebraic

equations. Note that the equation $\nabla \cdot \boldsymbol{B}' = 0$ gives no additional information to $-i\omega \boldsymbol{B} = \nabla \times (\boldsymbol{u} \times \boldsymbol{B})$ (take the divergence of both sides) so we drop $\nabla \cdot \boldsymbol{B}' = 0$ from further explicit consideration. Also p' can be substituted for in terms of ρ', using Eq. (5.16). The remaining 7 equations can be written schematically as $M\boldsymbol{x} = 0$ where $\boldsymbol{x} = (u_x, u_y, u_z, \rho', B_x', B_y', B_z')^T$ (superscript "T" denotes transpose) and the elements of matrix M are found from Eqs. (5.15). The condition for a non-trivial solution is that $\det M = 0$, which yields the dispersion relation

$$\left(\omega^2 - k_z^2 v_A^2\right) \left[\omega^4 - k^2 \left(c^2 + v_A^2\right)\omega^2 + k_z^2 k^2 v_A^2\right] = 0. \tag{5.17}$$

Waves for which $\omega^2 - k_z^2 v_A^2 = 0$ are called Alfvén waves. They have velocity and magnetic field perturbations perpendicular to both the equilibrium field direction and the wavenumber $\boldsymbol{k}$. Their phase speed $\omega/k = v_A |k_z|/k = v_A \cos\theta \leq v_A$ (θ being the angle between $\boldsymbol{k}$ and $\boldsymbol{B}$), with equality when $\boldsymbol{k}$ and $\boldsymbol{B}$ are parallel. Their group velocity is parallel (or antiparallel) to $\boldsymbol{B}$ and has magnitude v_A.

Waves for which $\omega^4 - k^2(c^2 + v_A^2)\omega^2 + k_z^2 k^2 v_A^2 = 0$ are called magnetoacoustic (or magnetosonic) waves. Evidently this condition is quadratic in ω^2 and so there are two solutions in ω^2: the corresponding waves are called the fast and slow magnetoacoustic waves. The phase speed along the field direction (ω/k_z) is between c_T and $\min(c, v_A)$ for the slow wave, and greater than $\max(c, v_A)$ for the fast wave. In the limit $v_A \to 0$, the slow wave frequencies tend to zero, while the fast waves become normal acoustic waves with $\omega^2 = k^2 c^2$. For further discussion of these waves and their physical characteristics, see e.g. Cowling (1976) and Roberts (1985).

We now consider briefly the case when the background field is unidirectional but *not* uniform, $\boldsymbol{B} = B(x)\boldsymbol{e}_z$ without loss of generality. Equilibrium (Eq. 5.14) requires $p(x) + B^2(x)/2\mu_0 = \text{constant}$. The Alfvén speed is also of course a function of x. We may seek solutions of the form $f(x)\exp(ik_y y + ik_z z - i\omega t)$ and, after some manipulation, Eqs. (5.15) and adiabatic energy equation (5.16) yield

$$\frac{\mathrm{d}}{\mathrm{d}x}\left(\frac{\rho(k_z^2 v_A^2 - \omega^2)(c^2 + v_A^2)(k_z^2 c_T^2 - \omega^2)}{(c^2 + v_A^2)(k_z^2 c_T^2 - \omega^2)k_y^2 + (k_z^2 c^2 - \omega^2)(k_z^2 v_A^2 - \omega^2)} \frac{\mathrm{d}v_x}{\mathrm{d}x}\right)$$
$$= \rho(x)\left(k_z^2 v_A^2(x) - \omega^2\right) v_x \tag{5.18}$$

(Chen & Hasegawa 1974; Roberts 1981, 1991). We shall not solve this here; but note that the differential equation has singularities where the coefficient of $\mathrm{d}^2 v_x/\mathrm{d}x^2$ vanishes: an Alfvén resonance at $\omega^2 = k_z^2 v_A^2(x)$

and a slow-mode/cusp resonance at $\omega^2 = k_z^2 c_T^2(x)$. These resonances occur at particular field lines, say $x = x_r$, and are called field-line resonances. They are important in the study of ultra low-frequency (ULF) waves in a planetary magnetosphere, and may also play a role in heating the Sun's corona (Section 5.3.3).

There has been much study of MHD surface waves at planar interfaces, and also of surface and body waves in magnetic flux tubes. See Roberts (1991) for a discussion in the context of MHD waves in the solar corona.

5.3   Some MHD Applications

We now consider some applications of MHD. It is appropriate that we draw on solar physics, since the Sun by its proximity provides a well observed example of magnetic fields in an astrophysical context.

5.3.1   *Solar prominences*

Prominences are long ribbons of cool, dense gas, situated in the hot rarified corona which forms the outer atmosphere of the Sun. They are long-lived structures, often persisting for many days. When viewed against the disk of the Sun they appear by contrast as dark lines and are called filaments. Prominences are located along neutral lines separating regions with opposite magnetic polarity in the photosphere (the Sun's visible surface).

The question of the origin of the denser material in the prominence is not completely settled, e.g. van Ballegooijen (2000). One possibility is that the material flows up from the lower atmosphere (chromosphere) in a syphon flow driven by pressure differences between the prominence and the chromosphere. Another possibility is that as magnetic flux ropes rise from the solar interior, arching through the photosphere and chromosphere, they carry material some of which remains in the prominence.

One simple model of a prominence, due to Kippenhahn & Schlüter (1957), is shown in Fig. 5.1. The relatively dense material in the prominence would fall under gravity unless supported by some other force. This opposing force is the Lorentz force, $j \times B$. We can explore the conditions required for such force balance using an extremely simple model in magnetostatic equilibrium (5.14).

Following Kippenhahn & Schlüter (1957), we model the prominence as a thin, vertical sheet in which the pressure $p(x)$, density $\rho(x)$ and vertical

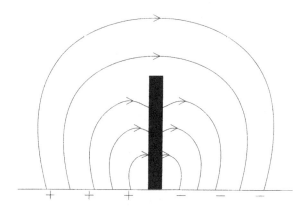

Fig. 5.1 Kippenhahn and Schlüter model of a prominence; the prominence material is shown as a shaded block.

magnetic field $B_z(x)$ depend only on horizontal coordinate x perpendicular to the sheet. The magnetic field components B_x and B_y are assumed to be constant, as is the temperature T and mean molecular weight μ (not to be confused with the magnetic permeability μ_0). Then the x and z components of Eq. (5.14) yield respectively

$$\frac{\mathrm{d}}{\mathrm{d}x}\left(p + \frac{B_z^2}{2\mu_0}\right) = 0 , \tag{5.19}$$

$$-\rho g + \frac{1}{\mu_0}B_x\frac{\mathrm{d}B_z}{\mathrm{d}x} = 0 . \tag{5.20}$$

We also require the equation of state relating pressure and density:

$$p = \frac{\mathcal{R}\rho T}{\mu} . \tag{5.21}$$

Equation (5.19) simply states that the gas pressure p adjusts to accommodate the magnetic pressure. Assuming that far from the sheet $(x \to \pm\infty)\ p \to 0$ and $B_z \to \pm B_{z\infty}$, Eq. (5.19) integrates to

$$p = \left(B_{z\infty}^2 - B_z^2\right)\Big/2\mu_0 . \tag{5.22}$$

Equation (5.20) states that the gravitational force is balanced by magnetic tension. This equation can be integrated, after eliminating density using Eqs. (5.21) and (5.22), to give

$$B_z(x) = B_{z\infty} \tanh\left(\frac{B_{z\infty}x}{2B_x H_p}\right) , \qquad (5.23)$$

where $H_p \equiv |dp/dz|^{-1}p = \mathcal{R}T/\mu g$ is the pressure scale height ($H_p \sim$ 200 km within the prominence). From Eqs. (5.22) and (5.23) we also obtain

$$p(x) = \frac{B_{z\infty}^2}{2\mu_0} \operatorname{sech}^2\left(\frac{B_{z\infty}x}{2B_x H_p}\right) . \qquad (5.24)$$

Equation (5.23) implies that the width of the prominence is of order $w \sim 2(B_x/B_{z\infty})H_p$. For observed prominences this implies ratios $B_{z\infty}/B_x$ of only a few tenths. Since for the field lines $dz/dx = B_z/B_x$, we deduce that only a slight depression of the field lines can produce enough tension to support the prominence.

The magnetic field configuration in real prominences is of course more complicated, with the prominence possibly being supported within a spiralled field: see e.g. van Ballegooijen (2000) for more discussion of prominence structures. However, these models can also be described using the same principles as used here.

5.3.2 Dynamo theory

The Sun, the Earth and most of the major planets have large-scale magnetic fields. The Earth has a mean field of about 0.3 G (1 Gauss $= 10^{-4}$ T, in SI units of Teslas). The Sun has a mean surface field of 1–2 G, though field strengths in sunspots are about 3000 G $= 0.3$ T. How are magnetic fields such as these generated and maintained? This question is the focus of dynamo theory. Developing realistic dynamo theories has proved very challenging indeed.

In the absence of fluid motions, the magnetic induction equation (5.7) reduces to a simple diffusion equation and implies that the magnetic field would decay on a timescale

$$t \sim \mathcal{L}^2 / \eta\pi^2 = \frac{\mu_0\sigma}{\pi^2}\mathcal{L}^2$$

where $\mathcal{L}$ is the lengthscale on which $\boldsymbol{B}$ varies. (This is derivable from dimensional analysis, except for the π^2 factor.) Taking $\mathcal{L}$ equal to 3500 km, the radius of the Earth's core, and $\eta \simeq 2.6$ m^2 s^{-1} gives a decay time for the Earth's field of 1–2$\times 10^4$ years. Yet rock magnetism indicates that the Earth has had a magnetic field for at least 10^9 years, requiring therefore that some dynamo regenerates the field. (In fact the Earth's dipole field polarity reverses on a timescale of order 10^4 years; any successful dynamo model for the Earth must also explain this result.)

For the Sun the decay timescale argument does not itself demand that a dynamo exist. Typical diffusion timescales quoted in the literature are 10^9–10^{10} years, comparable with or longer than the age of the Sun. Hence on this basis the field could be a remnant primordial field trapped and amplified geometrically by flux freezing of the weak (microgauss) Galactic field as the Sun collapsed from a gas cloud. However, the Sun's large-scale field reverses polarity on an approximately 11-year timescale. The sunspots and other indicators of magnetic activity also have such a cycle, with a quasi-period of about 22 years (taking into account the two polarities). It is generally believed that a dynamo is responsible for this behaviour.

How then might a dynamo work? The generic picture of dynamo action in a fluid contains the key ingredients of stretch, twist and fold. The magnetic field gets stretched out and intensified by the fluid motions, then twisted and folded back onto itself in an orientation that adds the field constructively. Finally a little diffusion "smoothes the joins". Geometrically an analogy is a closed loop of elastic bread dough, which can be stretched first to form a longer loop, then twisted into a figure of eight with a crossing in the middle, then one loop of the figure can be folded to lie on top of the other, and finally a bit of kneading together of the two loops and the twist leaves a single loop very much like the original dough loop.

An apparent impediment to dynamo theories, or rather to oversimplified attempts to model dynamo action, are the various so-called anti-dynamo theories. The first and best-known of these was established by Cowling (1934), who showed that a steady, axisymmetric magnetic field cannot be maintained by axisymmetric motions. Another such theorem states that it is impossible to maintain a 2-D field by dynamo action. Thus a successful dynamo must have at most a rather low degree of symmetry.

In our subsequent discussion we have in mind application to a star like the Sun with a substantial convection zone in which the fluid motion is turbulent.

The Sun is spherical to a very good approximation, and it is helpful first to introduce the decompositon of the magnetic field $\boldsymbol{B}$ into poloidal and toroidal components B_P and B_T:

$$\boldsymbol{B} \; = \; \boldsymbol{B}_P \; + \; \boldsymbol{B}_T \; .$$

In general, the definition of a toroidal field is one of the form $B_T = \nabla\times(\boldsymbol{r}f(\boldsymbol{r})) \equiv -\boldsymbol{r}\times\nabla f$, for some function $f(\boldsymbol{r})$; and the definition of a poloidal field is one of the form $B_P = \nabla\times(\nabla\times(\boldsymbol{r}f(\boldsymbol{r}))) \equiv -\nabla\times(\boldsymbol{r}\times\nabla f)$ for some other function f. It is straightforward to see that B_T has no radial component in spherical coordinates (r, θ, ϕ). If the system is axisymmetric, so $\partial/\partial\phi \equiv 0$, then B_T is purely in the azimuthal (ϕ) direction; and B_P has no ϕ-component and so B_P-lines lie wholly in meridional planes. (It may seem perverse to consider axisymmetric fields in the light of the anti-dynamo theorems, but the *mean* field might be axisymmetric even if in detail the field is not.)

We now discuss on a qualitative level how the large-scale field in the Sun might be generated. If at some initial time the Sun's field were purely poloidal then, unless the field lines were also lines of isorotation, differential fotation would stretch the lines out and wrap them in the azimuthal direction. Thus differential rotation can generate toroidal field B_T from a poloidal field B_P. This is sometimes called the ω-effect. The more difficult task to envisage is how poloidal field can be regenerated from the toroidal field, thus closing the cycle and establishing a dynamo. A possible mechanism was proposed by Parker (1955), who suggested that twisting, "cyclonic" motions in the convection could give toroidal field lines a twist locally, forming a loop which would then (if the field-line underwent say a quarter turn or three-quarter turn on average) be aligned in the poloidal direction and therefore contribute to a new B_P field. This is sometimes called the α-effect. A dynamo based on the above two ingredients is called an $\alpha\omega$-dynamo.

One framework in which to try to model the solar dynamo is the mean-field theory. This is described well in the book by Moffatt (1978). The basic idea is as follows. We suppose that the velocity of the flow can be written as a random fluctuating component $\boldsymbol{u}'(\boldsymbol{r}, t)$, which cannot itself be followed in detail, and a mean component $\bar{\boldsymbol{u}}(\boldsymbol{r}, t)$ which also varies with position and time but on longer length scales and time scales than does $\boldsymbol{u}'$. Similarly we suppose that the magnetic field can be decomposed into a mean component $\bar{\boldsymbol{B}}$ and a fluctuating component $\boldsymbol{B}'$. These can be substituted into the governing equations of the system, in particular into the induction equation (5.7). The means of fluctuating quantities individually are zero, but the means of products of fluctuating quantities will not be zero in general. In particular, in the induction equation the mean product $\nabla\times(\overline{\boldsymbol{u}'\times\boldsymbol{B}'})$ will contribute directly to the evolution of the mean field. This term looks like the curl of an electromotive force, and it can provide an α-

effect if the statistical flow-field lacks reflectional symmetry: see Moffatt (1978) for details.

We note as an aside that the decomposition into mean and fluctuating quantities can also be applied to the non-magnetic Navier-Stokes equation (1.9). Means of products of velocity fluctuations can then provide a force term on the mean flow of the form $\partial(-\overline{\rho u_i' u_j'})/\partial x_j$. Thus $(-\overline{\rho u_i' u_j'})$ has the effect of an additional component of the stress tensor as far as the mean flow $\bar{u}$ is concerned. It is called the Reynolds stress tensor.

Modern efforts to model the solar dynamo generally resort to large-scale computations. The problem of how the solar dynamo operates is not yet fully solved; but the above ideas, whilst probably not the full story, give insight into the likely ingredients. For further discussion, see e.g. Proctor & Gilbert (1994) and Proctor (2003). The most favoured site for the dynamo producing the Sun's large-scale field is the tachocline region at the base of the Sun's convection zone, though small-scale field may be generated in the bulk of the convection zone and particularly perhaps near the top. The problem of how flux tubes break away from the tachocline, rise through the convection zone and emerge through the photosphere to produce bipolar magnetic regions is also an interesting one: see Fisher *et al.* (2000) for a review relating theoretical ideas and observed properties of photospheric magnetic fields.

5.3.3 *Coronal heating*

The Sun's photosphere, its visible surface, is at a temperature of 5770 K. Paradoxically, as one goes up from the photosphere, into the rarified corona, the temperature increases rapidly to over 10^6 K. What heats the solar corona? It is not yet known for certain, though magnetic fields almost certainly play an essential role. Magnetic fields dominate the structure of the corona, as seen for example during a total solar eclipse. For a review of the coronal heating problem, see e.g. Priest (1993).

The heating requirements to maintain the corona at its observed temperature are fairly modest (compared to the Sun's total heat flux of $10^8 \, \mathrm{W\,m^{-2}}$): about $300 \, \mathrm{W\,m^{-2}}$ in quiet regions and coronal holes (where the coronal magnetic field lines open out into space) and $5000 \, \mathrm{W\,m^{-2}}$ above magnetically active regions; the problem is how to deposit the energy there. The electromagnetic energy flux $\boldsymbol{E} \times \boldsymbol{B}/\mu_0$ has magnitude $\sim |\boldsymbol{u}|\,|\boldsymbol{B}|^2/\mu_0 \sim 10^4 \, \mathrm{W\,m^{-2}}$ for velocities of order $100 \, \mathrm{m\,s^{-1}}$ and field strengths of $100 \, \mathrm{G} \, (= 10^{-2} \, \mathrm{T})$ is adequate, but must be converted into heat. In regions where dissipation is negligible, the flux-freezing approximation is valid and the field keeps its energy. What is required for heat deposition are singularities where the current density and gradient of B (and hence ohmic dissipation) are large. The principal candidates fall into two categories: wave heating and reconnection. Possibly both play a role.

Amongst the wave candidates, Alfvén and fast and slow magnetoacoustic waves have all been considered. Magnetoacoustic waves may dissipate by steepening to form shocks in the Sun's atmosphere, but are generally not thought capable of contributing more than a part of the heating required. More favoured candidates are Alfvén waves. These may dissipate by phase mixing (Heyvaerts & Priest 1983): waves initially excited in phase but travelling along neighbouring field lines corresponding to different Alfvén speeds eventually get sufficiently out of phase that strong gradients of field strength are set up and ohmic dissipation ensues. Alternatively, the waves may dissipate through the mechanism of resonant absorption (Ionson 1978): see Section 5.2.

The other major candidate is magnetic reconnection. When reconnection occurs, the frozen-field approximation breaks down locally and magnetic field lines reconnect, and magnetic energy is converted into heat and kinetic energy. This can occur when oppositely directed magnetic field lines are driven towards one another in a converging fluid flow. Ohmic dissipation takes place where the fields form a large field-strength gradient, and the field can reconnect to form a lower-energy topology. There are different proposals as to where reconnection takes place, but one possible site is when new, small-scale flux emerges from beneath the photosphere (generating the so-called magnetic carpet) and reconnects either with neighbouring small-scale flux or with existing overlying flux which it runs into as it rises. A comprehensive discussion of magnetic reconnection, together with application to the coronal heating problem, may be found in the book by Priest & Forbes (2000).

5.4   MHD Instabilities

In addition to the fluid dynamical instabilities considered in Chapter 4, new types of instability occur in magnetohydrodynamics, causing MHD equilibria to become unstable. These can lead to the formation of fine structure in the field, or sometimes to the occurence of energetic events such as solar flares or coronal mass ejections. Some common instabilities relevant to MHD applications have already been met in Chapter 4. These include the Rayleigh-Taylor instability and convective instability. New instabilities include the kink instability and sausage instability of magnetic flux tubes, when the tube is twisted too much, and the tearing-mode instability of a current sheet. Shear flows can lead to the Kelvin-Helmholtz instability which we have already met, and the Balbus-Hawley magneto-rotational instability in accretion disks.

Chapter 6

Numerical Computations

The complexity of the equations of fluid dynamics is such that, in general, it is not possible to solve them analytically. Thus it is commonly necessary to use numerical computational methods. Indeed, looking through almost any issue of the main astrophysical journals you will see the results of such numerical simulations. Computational fluid dynamics, even restricted to astrophysical applications, is a vast area and one we cannot hope to cover in one chapter. Thus we shall only consider some of the basics of computational fluid dynamics and hopefully give the reader a springboard for exploring the subject in the wider literature.

The reason that the area of computational fluid dynamics is so vast is that there is no one method that is best for all applications. It is a good rule to make as much analytical progress and simplification as possible before resorting to numerical methods, and then only to use as heavy a computational approach as is necessary. If, for example, the problem is one- or two-dimensional, it would be inefficient to solve the problem in three spatial dimensions. If the problem can adequately be addressed as a linearized perturbation about an equilibrium state, it would be unnecessary to follow the dynamical behaviour by solving numerically the full unlinearized equations.

We shall look mostly at finite differences, which are the basics of much numerical computation. However, there are other approaches to computing astrophysical fluid flows. We shall not even touch on spectral methods; but as an example of a very different approach we shall conclude by looking briefly at the formulation of Smoothed Particle Hydrodynamics (SPH).

6.1   The Formulation of Finite Differences

We consider as a simple illustrative example the numerical solution of the
1-D transport equation

$$\frac{\partial f}{\partial t} + u\frac{\partial f}{\partial x} = D\frac{\partial^2 f}{\partial x^2} \qquad (6.1)$$

for some quantity $f(x,t)$, where for simplicity we take the advection velocity
u and diffusion coefficient D to be non-negative constants. We assume that
we are evolving f forward in time starting with given initial conditions,
and with spatial boundary conditions prescribed. A widely used numerical
approach is to represent f on a discrete grid

$$x_0, x_1, \ldots, x_N \qquad (x_0 < x_1 < x_2 < \ldots < x_N).$$

We shall assume a uniform grid spacing Δx, so $x_{j+1} - x_j = \Delta x$, but what
follows can be generalized for a non-uniform spacing. Also we march the
solution forward from initial time t_0 with a timestep Δt, computing the
solution at times

$$t_1 = t_0 + \Delta t, \, t_2 = t_1 + \Delta t = t_0 + 2\Delta t, \, \ldots, \, t_n = t_0 + n\Delta t, \, \ldots.$$

Let f_j^n denote our numerical solution for f at $x = x_j$, $t = t_n$. The idea
is, given the solution at all meshpoints at time t_n – and, if we wish, at
previous times also, though this then requires additionally storing in the
computer – to compute the solution at the next time t_{n+1}.

There are various ways of forming discretized approximations to the
derivatives in Eq. (6.1). For example, $\partial f/\partial x$ at $x = x_j$ and $t = t_n$ may be
approximated variously as

$$\left.\frac{\partial f}{\partial x}\right|_j^n \simeq (f_{j+1}^n - f_j^n)/\Delta x \qquad \text{forward differences}$$

$$\left.\frac{\partial f}{\partial x}\right|_j^n \simeq (f_j^n - f_{j-1}^n)/\Delta x \qquad \text{backward differences} \qquad (6.2)$$

$$\left.\frac{\partial f}{\partial x}\right|_j^n \simeq (f_{j+1}^n - f_{j-1}^n)/2\Delta x \qquad \text{centred differences}.$$

Of course these are only approximations to the derivatives, even if the values
f_j^n etc. are known precisely. To estimate the order of the error, in terms of

grid spacing Δx, we can use the Taylor series for f:

$$f(x + h, t) = f(x, t) + h\frac{\partial f}{\partial x} + \frac{h^2}{2!}\frac{\partial^2 f}{\partial x^2} + \frac{h^3}{3!}\frac{\partial^3 f}{\partial x^3} + O(h^4). \quad (6.3)$$

Putting $x = x_j$, $h = \Delta x$ in Eq. (6.3), we find that the leading error in the forward-difference approximation to $\partial f/\partial x$ is $\frac{1}{2}(\Delta x)\partial^2 f/\partial x^2$, i.e. the error is $O(\Delta x)$; likewise for backward differences. For centred differences, the error is $O(\Delta x^2)$ because the $\partial^2 f/\partial x^2$ terms cancel.

To estimate the first derivative requires values of f at (at least) two points; to estimate higher derivatives requires more. For example, we can approximate the second derivative $\partial^2 f/\partial x^2$ in Eq. (6.1) at $x = x_j$, $t = t_n$ as

$$\left.\frac{\partial^2 f}{\partial x^2}\right|_j^n \simeq \left(f_{j+1}^n - 2f_j^n + f_{j-1}^n\right)/(\Delta x)^2 \quad (6.4)$$

with three adjacent points. Such discrete approximations, including higher-order formulae and formulae appropriate to non-uniform grids, can be derived by fitting polynomials through adjacent points (x_k, f_k^n) and evaluating the required derivative of the polynomial at $x = x_j$. Alternatively, they can be derived using the Taylor series (6.3): the Taylor series also yields the order of the error in the approximation.

The time derivative in Eq. (6.1) can be approximated in like manner. The most obvious way is to use forward differences, since then the values at t_{n+1} are computed directly from values at t_n. A discretized approximation to Eq. (6.1) is then

$$\frac{\left(f_j^{n+1} - f_j^n\right)}{\Delta t} = -u\frac{\left(f_{j+1}^n - f_{j-1}^n\right)}{2\Delta x} + D\frac{\left(f_{j+1}^n - 2f_j^n + f_{j-1}^n\right)}{(\Delta x)^2} \quad (6.5)$$

where centred differences have been used for the advection term. We may write Eq. (6.5) as

$$f_j^{n+1} = f_j^n - \frac{1}{2}\alpha\left(f_{j+1}^n - f_{j-1}^n\right) + \beta\left(f_{j+1}^n - 2f_j^n + f_{j-1}^n\right) \quad (6.6)$$

where $\alpha = u\Delta t/\Delta x$, $\beta = D\Delta t/(\Delta x)^2$.

6.2   The von Neumann Stability Analysis

A vital issue of practical importance is whether the discretized scheme is *stable*, or whether the solution will become progressively dominated by error with increasing time. A useful tool for assessing stability is the von Neumann stability analysis. Since Eq. (6.6) is linear in f and with constant coefficients, it admits independent solutions

$$\xi^n e^{ikj\Delta x} \qquad (6.7)$$

where k is a real spatial wavenumber (which can take any value), and $\xi(k)$ is an amplitude which depends on k. Clearly, $|f_j^{n+1}|/|f_j^n| = |\xi|$, so if $|\xi| > 1$ the modulus of the solution will grow with time and tend to infinity as $n \to \infty$. In this case, the scheme is *unstable*, since any small perturbation of our solution by such an unstable mode will grow without bound. Thus a condition for stability is that the *amplification factor* ξ satisfies $|\xi| \le 1$ for all k.

Substituting expression (6.7) into Eq. (6.6) and dividing by f_j^n gives

$$\begin{aligned}
\xi &= 1 - \frac{1}{2}\alpha\left(e^{ik\Delta x} - e^{-ik\Delta x}\right) + \beta\left(e^{ik\Delta x} - 2 + e^{-ik\Delta x}\right) \\
&= 1 - i\alpha\sin k\Delta x + 2\beta\left(\cos k\Delta x - 1\right)
\end{aligned} \qquad (6.8)$$

so that

$$|\xi|^2 = \left(1 - 4\beta\sin^2\frac{1}{2}k\Delta x\right)^2 + \alpha^2\sin^2 k\Delta x . \qquad (6.9)$$

Hence the condition for stability is that

$$\left(1 - 4\beta\sin^2\frac{1}{2}k\Delta x\right)^2 + \alpha^2\sin^2 k\Delta x \le 1 \qquad (6.10)$$

for all $k\Delta x$. This is equivalent to

$$0 \le \alpha^2 \le 2\beta \le 1 . \qquad (6.11)$$

This imposes conditions on the timestep and grid spacing, depending on the values of u and D.

If $\alpha = 0$ (no advection), the stability criterion is that $2\beta \le 1$, i.e. $2D\Delta t/(\Delta x)^2 \le 1$. That is, the timestep must not exceed the diffusion time across one mesh spacing.

If instead $\beta = 0$ (no diffusion), however, then conditions (6.10), (6.11) cannot be satisfied, so for the non-diffusive problem the scheme is unstable no matter how small a timestep is chosen. Fortunately there are various alternatives to (6.6) which make the non-diffusive case stable, as we shall now discuss.

6.3 Various Finite-difference Schemes

6.3.1 *The Lax method*

As an alternative to (6.6), one can replace f_j^n in the time derivative by the average $(f_{j+1}^n + f_{j-1}^n)/2$ of the values at the two neighbouring points. Then, with $\beta = 0$, Eq. (6.6) becomes

$$f_j^{n+1} = \frac{1}{2}\left(f_{j+1}^n + f_{j-1}^n\right) - \frac{1}{2}\alpha\left(f_{j+1}^n - f_{j-1}^n\right) . \tag{6.12}$$

This is the Lax method.

The von Neumann analysis gives

$$\xi = \cos k\Delta x - i\alpha \sin k\Delta x$$

so $|\xi|^2 = \cos^2 k\Delta x + \alpha^2 \sin^2 k\Delta x \leq 1$ for all $k\Delta x$ provided $\alpha \leq 1$. That is, for stability we require

$$u\Delta t/\Delta x \leq 1 . \tag{6.13}$$

This is the well-known *Courant stability criterion* (or Courant-Friedrichs-Lewy criterion). Physically, the timestep must be smaller than the advection time across one mesh spacing.

One can re-arrange (6.12) as

$$\frac{\left(f_j^{n+1} - f_j^n\right)}{\Delta t} = -u\frac{\left(f_{j+1}^n - f_{j-1}^n\right)}{2\Delta x} + \frac{(\Delta x)^2}{2\Delta t}\frac{\left(f_{j+1}^n - 2f_j^n + f_{j-1}^n\right)}{(\Delta x)^2} . \tag{6.14}$$

Comparing this with Eq. (6.5) shows that the adjustment in Lax's method is equivalent to introducing a diffusive term with diffusion coefficient $\Delta x^2/2\Delta t$: this is sufficient to damp out the unstable behaviour.

6.3.2 *Upwind differencing*

Another alternative to (6.6) with $\beta = 0$ is to use an *upwind scheme*, i.e. approximate the derivative in the advection term with a backward difference (for positive u):

$$\left(f_j^{n+1} - f_j^n\right)/\Delta t = -u\left(f_j^n - f_{j-1}^n\right)/\Delta x . \tag{6.15}$$

The von Neumann stability analysis of this scheme yields

$$\xi = 1 - \alpha\left(1 - e^{-ik\Delta x}\right) .$$

The stability condition $|\xi|^2 \leq 1$ again leads to the Courant criterion (6.13). By a similar manipulation to that which led to (6.14), one can show that the upwind scheme is equivalent to introducing a diffusive term, with diffusion coefficient $u\Delta x/2$.

6.3.3 The staggered leapfrog method

Thus far we have considered only discretizations that are first-order accurate in time. It is of course possible to take second- or higher-order time approximations, so for example

$$f_j^{n+1} - f_j^{n-1} = \frac{u\Delta t}{\Delta x} \left(f_{j+1}^n - f_{j-1}^n \right) \qquad (6.16)$$

uses centred differences for both the time derivative and the advection term. It is called the *staggered leapfrog method*, because the time levels in the time derivative "leapfrog" over the time levels in the spatial derivative. The method requires that the solution be stored at the two previous times t_n and t_{n-1} in order to calculate quantities at t_{n+1}. A von Neumann analysis leads to a quadratic equation in the amplification factor: for stability the Courant condition must again be satisfied. A problem however with such a scheme is that quantities at space-time points (x_j, t_n) where $j + n$ is even do not communicate with quantities at points where $j + n$ is odd, and this leads to an instability called mesh drift (see Press *et al.* 1989).

6.3.4 The Lax-Wendroff method

A second-order method in time that avoids the problem of mesh drift is the two-step Lax-Wendroff method (e.g. Press *et al.* (1989)). Considering again our non-diffusive advection problem, one defines quantities $f_{j+\frac{1}{2}}^{n+\frac{1}{2}}$ at half timesteps $t_{n+\frac{1}{2}}$ and half mesh points $x_{j+\frac{1}{2}}$. The first, Lax, step calculates these via

$$\left[f_{j+\frac{1}{2}}^{n+\frac{1}{2}} - \frac{1}{2} \left(f_{j+1}^n + f_j^n \right) \right] / \left(\frac{1}{2}\Delta t \right) = -u \left(f_{j+1}^n - f_j^n \right) / \Delta x . \qquad (6.17)$$

In the second step, the quantities at t_{n+1} are calculated using a centred expression

$$\left(f_j^{n+1} - f_j^n \right) / \Delta t = -u \left(f_{j+\frac{1}{2}}^{n+\frac{1}{2}} - f_{j-\frac{1}{2}}^{n+\frac{1}{2}} \right) / \Delta x . \qquad (6.18)$$

In the Lax-Wendroff method, the values $f_{j+\frac{1}{2}}^{n+\frac{1}{2}}$ are no longer retained after the step has been taken.

6.3.5   *Implicit schemes: the Crank-Nicholson method*

Equation (6.5) gives an *explicit scheme*: the solution f_j^{n+1} at the $(n+1)^{\text{th}}$ timestep is given explicitly in terms of the solutions already obtained at earlier times. As applied, for example, to the pure diffusion problem, it requires timesteps such that $2\beta \leq 1$, i.e. $2D\Delta t/(\Delta x)^2 \leq 1$. If we are primarily interested in the evolution on lengthscales $L \gg \Delta x$, the number of timesteps required to see such evolution is of order $L^2/(\Delta x)^2$ (independent of D), which can be prohibitively expensive computationally. An alternative approach for the purely diffusive problem is

$$\left(f_j^{n+1} - f_j^n\right)/\Delta t \;=\; D\left(f_{j+1}^{n+1} - 2f_j^{n+1} + f_{j-1}^{n+1}\right)/\Delta x^2 \tag{6.19}$$

where now the values on the right-hand side are evaluated at time t_{n+1}. This is an example of a fully *implicit* scheme. The solution f_j^{n+1} is now not quite so immediate; but (6.19) can be re-arranged as

$$-\beta f_{j+1}^{n+1} + (1+2\beta)f_j^{n+1} - \beta f_j^{n-1} \;=\; f_j^n \tag{6.20}$$

so that with appropriate boundary conditions the solution at t_{n+1} is now found by inverting a tridiagonal matrix (e.g. Press *et al.* (1989) for methods of doing this). A von Neumann stability analysis of (6.20) gives amplification factor

$$\xi \;=\; \frac{1}{1 + 4\beta \sin^2 \frac{1}{2}k\Delta x}$$

which shows that the implicit scheme is stable for any value of Δt, no matter how large. Stability does not imply accuracy, however, and the small-scale evolution will not be accurately followed when Δt is large. Moreover, Eq. (6.19) is only first-order accurate in time. Another approach is to replace the diffusive operator by the average of the explicit and implicit representations:

$$\left(f_j^{n+1} - f_j^n\right)/\Delta t \;=\; \frac{D}{\Delta x^2}\left(\frac{1}{2}\partial_2^{n+1}f + \frac{1}{2}\partial_2^n f\right) \tag{6.21}$$

where $\partial_2^{n+1}f \equiv f_{j+1}^{n+1} - 2f_j^{n+1} + f_{j-1}^{n+1}$ and $\partial_2^n f \equiv f_{j+1}^n - 2f_j^n + f_{j-1}^n$. This is the Crank-Nicholson scheme. Like the implicit scheme it is unconditionally stable for all Δt; but it is also second-order accurate in time. Thus not only is the scheme stable with large timesteps, it also produces higher accuracy.

To see why Crank-Nicholson is second-order accurate in time, note first that the right-hand side of Eq. (6.21) can be written as

$$\frac{D}{(\Delta x)^2} \left(f_{j+1}^{n+\frac{1}{2}} - 2f_j^{n+\frac{1}{2}} + f_{j-1}^{n+\frac{1}{2}} \right)$$

where, for any j, $f_j^{n+\frac{1}{2}} \equiv (f_j^{n+1} + f_j^n)/2$. This approximates the second derivative $\partial^2 f/\partial x^2$ at time $t_{n+\frac{1}{2}} \equiv t_n + \frac{1}{2}\Delta t$: the error in this approximation is $O(\Delta x)^2$. Now we can interpret the left-hand side of (6.21) as an approximation (to $\partial f/\partial t$) also at $t_{n+\frac{1}{2}}$. By expanding each of f_j^{n+1} and f_j^n in turn as a Taylor series in time about $t_{n+\frac{1}{2}}$, and subtracting one from the other, it follows that the left-hand side of Eq. (6.21) is

$$(\partial f/\partial t)_j^{n+\frac{1}{2}} + O(\Delta t)^2 \tag{6.22}$$

by symmetry. Thus the scheme is second-order accurate in time and (from above) second-order accurate in space.

Crank-Nicholson is a good choice of scheme for simple diffusion problems.

The scheme in Eq. (6.21) can be generalized to

$$\left(f_j^{n+1} - f_j^n \right)/\Delta t = \frac{D}{\Delta x^2} \left(\theta \partial_2^{n+1} f + (1-\theta)\partial_2^n f \right) \tag{6.23}$$

where θ is a parameter. In particular, $\theta = 0, 1$ give the explicit and implicit schemes, respectively, while $\theta = 1/2$ yields the Crank-Nicholson scheme.

6.4   Considerations for More Complex Systems

The von Neumann analysis may be extended to cope with coupled equations such as we must often solve in fluid dynamics. If the M difference equations can be written schematically for example as

$$\boldsymbol{y}_j^{n+1} = A_+ \boldsymbol{y}_{j+1}^n + A_0 \boldsymbol{y}_j^n + A_- \boldsymbol{y}_{j-1}^n \tag{6.24}$$

where $\boldsymbol{y}$ is an M-vector and the As are constant matrices, this has independent solutions

$$\boldsymbol{y}_j^n = \xi^n e^{ikj\Delta x} \boldsymbol{y}_0 \tag{6.25}$$

with $\boldsymbol{y}_0$ a constant vector. Substituting (6.25) into (6.24) shows that ξ and $\boldsymbol{y}_0$ are respectively eigenvalues and eigenvectors of $A(k) \equiv e^{ik\Delta x}A_+ + A_0 + e^{-ik\Delta x}A_-$: the condition for stability is that $|\xi| \le 1$ for all eigenvalues of A.

So far we have supposed that the coefficients in our differential equations are constants, but generally they will vary in space (and possibly in time) or be functions of the variables being solved for. Several brief remarks are in order. Firstly, with regard to the von Neumann analysis, even if the coefficients vary they are treated in the stability analysis as constants, the assumption being that they vary slowly compared with the solution being studied. If the coefficients depend on the independent variable (f), so the problem is non-linear, one can write f as $f_0 + \delta f$ and linearize in δf: assuming the f_0 terms already satisfy the difference equations exactly, the stability analysis would look for an unstable eigenmode for δf. If for example D varies with x, terms such as $D\partial^2 f/\partial x^2$ in Eq. (6.5) can be discretized as

$$\left[D_{j+\frac{1}{2}}\left(f_{j+1}^n - f_j^n\right) - D_{j-\frac{1}{2}}\left(f_j^n - f_{j-1}^n\right) - \right]/\Delta x^2 \qquad (6.26)$$

where $D_{j\pm\frac{1}{2}} = D(x_{j\pm\frac{1}{2}})$. If $D = D(f)$, then an explicit scheme can proceed in a similar fashion with $D_{j+\frac{1}{2}} = [D(f_{j+1}^n) + D(f_j^n)]/2$, etc. Implicit schemes are not so easy to generalize, but non-linear difference equations can be avoided by linearizing e.g. $D(f_j^{n+1})$ as $D(f_j^n) + (f_j^{n+1} - f_j^n)\left.\frac{\partial D}{\partial f}\right|_j^n$. When the advection velocity u is a variable, say u_j^n, an upwind scheme such as (6.15) must be implemented in such a way as to take account of whether u_j^n is positive or negative.

6.5 Operator Splitting

A concept widely used in modern finite-difference schemes in computational fluid dynamics is *operator splitting*. Suppose that the initial-value problem to be solved is

$$\frac{\partial f}{\partial t} = \mathcal{L}f \qquad (6.27)$$

where $\mathcal{L}$ is some operator. Suppose further that $\mathcal{L}$ can be written as the sum of m operators

$$\mathcal{L} = \mathcal{L}_1 + \mathcal{L}_2 + \ldots + \mathcal{L}_m ,$$

for each of which we already have a finite-difference scheme for updating f from time t_n to time t_{n+1}, so $f^{n+1} = U_p(f^n)$ if the problem to be solved were $\partial f/\partial t = \mathcal{L}_p f$, for each of $p = 1, 2, \ldots, m$ separately. A form of operator splitting is to compute the solution f^{n+1} to (6.27) from f^n in m fractional steps thus:

$$
\begin{aligned}
f^{n+1/m} &= U_1\left(f^n, \Delta t\right) \\
f^{n+2/m} &= U_2\left(f^{n+1/m}, \Delta t\right) \\
&\ldots \\
f^{n+1} &= U_m\left(f^{n+(m-1)/m}, \Delta t\right).
\end{aligned}
\tag{6.28}
$$

For example, the combined advection-diffusion problem (6.1) could be split into two operators, $\mathcal{L}_1$ being the advection operator and $\mathcal{L}_2$ the diffusion operator; and U_1 could be an explicit scheme for the advection term while U_2 could be the Crank-Nicholson scheme for the diffusion operator.

A different kind of operator splitting can be used, in which each U_p is an update scheme for the whole operator $\mathcal{L}$ but which may only be stable for the $\mathcal{L}_p$ part of the operator. The implementation is similar to above, but now each update takes a fractional timestep $\Delta t/m$, not the full timestep Δt:

$$
\begin{aligned}
f^{n+1/m} &= U_1\left(f^n, \Delta t/m\right) \\
f^{n+2/m} &= U_2\left(f^{n+1/m}, \Delta t/m\right) \\
&\ldots \\
f^{n+1} &= U_m\left(f^{n+(m-1)/m}, \Delta t/m\right).
\end{aligned}
\tag{6.29}
$$

An example is the Alternating Direction Implicit (ADI) method. For the two-dimensional diffusion problem

$$
\frac{\partial f}{\partial t} = D\left(\frac{\partial^2 f}{\partial x^2} + \frac{\partial^2 f}{\partial y^2}\right),
\tag{6.30}
$$

for instance, ADI has $m = 2$ and in the first update (U_1) ADI uses the implicit scheme for $\partial^2/\partial x^2$ and the explicit scheme for $\partial^2/\partial y^2$; and conversely for the second update (U_2). Each substep still requires only the inversion of a tri-diagonal matrix. By contrast, for such multi-dimensional diffusion problems the Crank-Nicholson scheme loses its tri-diagonal character, resulting in a computationally more expensive matrix inversion. See Press *et al.* (1989) for more details.

As discussed above – cf. Eqs. (6.24), (6.25) – the above principles and various techniques can be applied to coupled equations as are encountered in astrophysical fluid dynamics. The dependent variable f in e.g. Eq. (6.27) can then be interpreted as a vector of dependent variables, e.g. three components of velocity, density, internal energy. These differential equations must generally be supplemented by an equation of state which must also be imposed.

6.6 Examples of Implementations

To show how finite-difference calculations may be implemented in astrophysical fluid dynamics, we now consider two examples taken from the literature. The first is for 1-D flow in a gravitationally stratified medium: this is analogous to radial motion in a stellar atmosphere, and can indeed be extended to model that, though here we restrict attention to Cartesian geometry only. The second example is for 2-D flow (rotation and circulation) in the radiative interior of the Sun. We shall conclude with some comments about more sophisticated implementations to solving astrophysical fluid-flow problems.

6.6.1 *1-D Lagrangian scheme with artificial viscosity*

In this example we set up a set of finite-difference equations for solving for 1-D flow in a gravitationally stratified medium in planar geometry. We follow closely the presentation by Mihalas & Mihalas (1984). When considering flow in such a system, as also in radial oscillations of a star, it is convenient to express the equations in terms of a Lagrangian mass coordinate m rather than a spatial coordinate r (or in our planar case, z). The reason is that in many cases the temporal variability evaluated at fixed m is smaller than the variability evaluated at a fixed point in space. Hence we define a mass coordinate (mass per unit area):

$$m(z) = \int_0^z \rho(z')\,\mathrm{d}z'\,. \tag{6.31}$$

(For the sake of definiteness we suppose that z and m increase in the direction opposite to the gravitational acceleration.) Fluid quantities are now considered to be functions of m and t. The derivative with respect to t at fixed m is just the material derivative D/Dt. The equations to be solved are

$$(Du/Dt) = -(\partial p/\partial m) - g \qquad (6.32)$$

$$(Dz/Dt) = u \qquad (6.33)$$

$$V \equiv 1/\rho = (\partial z/\partial m) \qquad (6.34)$$

$$(DU/Dt) = -p(DV/Dt) + \dot{q} \qquad (6.35)$$

where V is specific volume and $\dot{q}$ represents possible energy sources or losses.

Before proceeding to solve these equations, we introduce the concept of artificial viscosity. The topic of shocks in astrophysical fluid flow is discussed in Chapter 8. However, at this stage we note that in order to handle shocks in the difference equations it is expedient to introduce an *artificial viscosity*. This is achieved by replacing p in the momentum and energy equations by $p + \Pi$, where Π is an equivalent viscous pressure. We aim to avoid unresolved discontinuous shocks in our solutions by spreading shocks over several mesh points. This could be achieved simply by using a large constant viscosity, but this would affect the flow everywhere. Instead, von Neumann and Richtmyer proposed using an artificial viscosity that is large in shocks but very small elsewhere. Hence we define

$$\Pi = \begin{cases} \frac{4}{3}\rho^{-1}l^2(D\rho/Dt)^2 & \text{where } D\rho/Dt > 0, \text{ i.e., } \partial u/\partial z < 0 \\ 0 & \text{elsewhere} \end{cases} . \qquad (6.36)$$

The parameter l has dimensions of length and is typically taken to be a small multiple of the mesh spacing: $l = k\Delta z$ where k is between 1.5 and 2 typically.

The equations are solved using a staggered mesh, with velocities evaluated at integer spatial points and half-integer time steps, and pressure and density are evaluated at integer time steps and half-integer spatial points (Fig. 6.1). An explicit scheme for the momentum equation is

$$\left(u_j^{n+\frac{1}{2}} - u_j^{n-\frac{1}{2}}\right)/\Delta t = -g - \left(p_{j+\frac{1}{2}}^n - p_{j-\frac{1}{2}}^n + \Pi_{j+\frac{1}{2}}^{n-\frac{1}{2}} - \Pi_{j-\frac{1}{2}}^{n-\frac{1}{2}}\right)/\Delta m . \qquad (6.37)$$

The artificial viscosity is computed as

$$\Pi_{j+\frac{1}{2}}^{n-\frac{1}{2}} = k^2 \frac{\rho_{j+\frac{1}{2}}^n + \rho_{j+\frac{1}{2}}^{n-1}}{2} \left(u_{j+1}^{n-\frac{1}{2}} - u_j^{n-\frac{1}{2}}\right)^2 . \qquad (6.38)$$

(For formulae appropriate to non-uniform step-sizes, see Mihalas & Mihalas 1984.) Note in (6.36) that Π is evaluated at $t_{n-\frac{1}{2}}$ rather than at t_n; but this deficiency does not generally produce large errors.

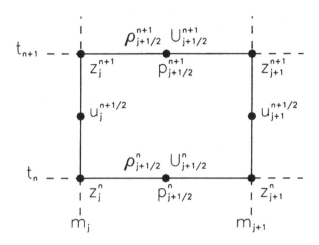

Fig. 6.1 The staggered mesh used for the 1-D Lagrangian example. Thermodynamic quantities (pressure p, density ρ, internal energy U) are computed at half-integer spatial points. Velocities are computed at half-integer time steps.

From the velocities, z_j^{n+1} is calculated as

$$z_j^{n+1} = z_j^n + u_j^{n+\frac{1}{2}} \Delta t \qquad (6.39)$$

and then the specific volumes are updated:

$$V_{j+\frac{1}{2}}^{n+1} = 1/\rho_{j+\frac{1}{2}}^{n+1} = \left(z_{j+1}^{n+1} - z_j^{n+1} \right) / \Delta m . \qquad (6.40)$$

Thence Π at $t_{n+\frac{1}{2}}$ can be evaluated.

The energy equation is solved last, using an implicit scheme

$$U_{j+\frac{1}{2}}^{n+1} - U_{j+\frac{1}{2}}^n + \left(\frac{p_{j+\frac{1}{2}}^n + p_{j+\frac{1}{2}}^{n+1}}{2} + \Pi_{j+\frac{1}{2}}^{n+\frac{1}{2}} \right) \left(V_{j+\frac{1}{2}}^{n+1} - V_{j+\frac{1}{2}}^n \right)$$

$$= \Delta t \dot{q}_{j+\frac{1}{2}}^{n+\frac{1}{2}} \qquad (6.41)$$

where $\dot{q}_{j+\frac{1}{2}}^{n+\frac{1}{2}} \equiv (1-\theta)\dot{q}_{j+\frac{1}{2}}^n + \theta\dot{q}_{j+\frac{1}{2}}^{n+1}$. Parameter θ can be set to $\theta = \frac{1}{2}$ (cf. Crank-Nicholson) or to $\theta = 1$ for a fully implicit evaluation. A remaining problem is that $p_{j+\frac{1}{2}}^{n+1}$ is not known. If, for example, p, U and $\dot{q}$ are known functions of ρ and temperature T, then by making an estimate T^* for T^{n+1} the pressure, internal energy and $\dot{q}$ can be linearized around $T = T^*$. The implicit equation (6.41) can then be solved and the estimate T^* improved. The process can be iterated until the updates to T^* are acceptably small. More details are given by Mihalas & Mihalas (1984).

Finally, a few words about applying boundary conditions. A prescribed velocity $F(t)$ at one end (for convenience we consider the end $j = 0$), as for example if the end is driven by a piston, can be imposed as $u_0^{n+\frac{1}{2}} = F(t^{n+\frac{1}{2}})$. A free boundary can be imposed as $u_1^{n+\frac{1}{2}} = u_0^{n+\frac{1}{2}}$. For zero total surface pressure ($p + \Pi = 0$ at $z = z_0$) an appropriate boundary condition is

$$u_0^{n+\frac{1}{2}} - u_0^{n-\frac{1}{2}} = -g\Delta t - \left(p_{\frac{1}{2}}^n + \Pi_{\frac{1}{2}}^{n-\frac{1}{2}}\right) \Delta t / (\frac{1}{2}\Delta m) . \qquad (6.42)$$

For the results of a stability analysis of this system, see Mihalas & Mihalas (1984) and Richtmyer & Morton (1967). In brief, outside any shocks the usual Courant condition holds. With shocks, reducing the timestep by about a factor of two is usually sufficient to provide stability. As Mihalas & Mihalas (1984) comment, to ensure accuracy as well as stability one may choose to limit the timesteps so that in a single timestep the fractional change in any variable is less than some chosen upper bound.

6.6.2   *2-D scheme using operator splitting*

In this example we discuss a finite-difference approach to solving for 2-D flow in the radiative interior of the Sun or solar-like star. We follow the presentation by Elliott (1996). Assuming axisymmetry, the fluid equations are expressed in spherical polar coordinates as

$$\frac{\partial u_r}{\partial t} = -u_r \frac{\partial u_r}{\partial r} - \frac{u_\theta}{r} \frac{\partial u_r}{\partial \theta} - \frac{1}{\rho} \frac{\partial p}{\partial r} - g + r\Omega^2 \sin^2 \theta$$
$$+ \nu \left(\nabla^2 u_r - \frac{2}{r^2} u_r - \frac{2}{r^2} \frac{\partial u_\theta}{\partial \theta} - \frac{2}{r^2} \frac{\cos \theta}{\sin \theta} u_\theta \right) \qquad (6.43)$$

$$\frac{\partial u_\theta}{\partial t} = -u_r \frac{\partial u_\theta}{\partial r} - \frac{u_\theta}{r} \frac{\partial u_\theta}{\partial \theta} - \frac{1}{\rho r} \frac{\partial p}{\partial \theta} + r\Omega^2 \sin \theta \cos \theta$$
$$+ \nu \left(\nabla^2 u_\theta - \frac{1}{r^2 \sin^2 \theta} u_\theta + \frac{2}{r^2} \frac{\partial u_r}{\partial \theta} \right) \qquad (6.44)$$

$$\frac{\partial u_\phi}{\partial t} = -u_r \frac{\partial u_\phi}{\partial r} - \frac{u_\theta}{r} \frac{\partial u_\phi}{\partial \theta} - 2\Omega \left(u_r \sin\theta + u_\theta \cos\theta \right)$$

$$+ \nu \left(\nabla^2 u_\phi - \frac{1}{r^2 \sin^2\theta} u_\phi \right) \tag{6.45}$$

$$\frac{\partial p}{\partial t} = -u_r \frac{\partial p}{\partial r} - \frac{u_\theta}{r} \frac{\partial p}{\partial \theta} - \gamma p \left[\frac{1}{r^2} \frac{\partial}{\partial r} \left(r^2 u_r \right) + \frac{1}{r \sin\theta} \frac{\partial}{\partial \theta} \left(u_\theta \sin\theta \right) \right]$$

$$+ (\gamma - 1) \left[\rho \epsilon + \frac{1}{r^2} \frac{\partial}{\partial r} \left(r^2 K \frac{\partial T}{\partial r} \right) + \frac{1}{r^2 \sin\theta} \frac{\partial}{\partial \theta} \left(K \sin\theta \frac{\partial T}{\partial \theta} \right) \right] \tag{6.46}$$

$$\frac{\partial \rho}{\partial t} = -u_r \frac{\partial \rho}{\partial r} - \frac{u_\theta}{r} \frac{\partial \rho}{\partial \theta} - \rho \left[\frac{1}{r^2} \frac{\partial}{\partial r} \left(r^2 u_r \right) + \frac{1}{r \sin\theta} \frac{\partial}{\partial \theta} \left(u_\theta \sin\theta \right) \right] . \tag{6.47}$$

Here $\Omega = u_\phi / r \sin\theta$ is the angular velocity; the energy generation term ϵ includes any contributions from e.g. viscous dissipation; and K is the radiative conductivity ($K = 4acT^3 / 3\kappa\rho$).

Although we do not go into full details here, Elliott (1996) makes a transformation to take account of the oblateness, by defining new coordinates $r' = r + f(r', \theta')$ and $\theta' = \theta$, the function f being chosen so that the modified potential (3.8) for the initial angular rotation rate is constant on surfaces of constant r'. The partial differential operators $\partial/\partial r$ and $\partial/\partial\theta$ become $\partial/\partial r'$ and $\partial/\partial\theta' + (\partial r'/\partial\theta)\partial/\partial r'$ respectively. These are substituted into the above equations, and only terms up to $O(\Omega^4)$ retained, assuming that K and η may be of order unity. Finally the primes on the coordinates may be dropped. The only places where the equations are affected are in the $\partial p/\partial\theta$ and $\partial T/\partial\theta$ terms.

The system of equations is solved using the ADI method (Section 6.5). Terms involving derivatives with respect to r are put into one operator ($\mathcal{L}_r$) and those involving derivatives with respect to θ put into another ($\mathcal{L}_\theta$). The remaining terms are split between $\mathcal{L}_r$ and $\mathcal{L}_\theta$. These two operators are then discretized by replacing derivatives using centred differences. In this way a scheme which is second order in both space and time is obtained.

The timestep must be chosen to ensure stability and to avoid excessive truncation errors due to the discretization in time. The first requirement is not very stringent for a semi-implicit scheme like ADI. The second essentially means that the time discretization should not introduce larger truncation errors than the spatial discretization. This is satisfied provided fluid elements do not travel more than one mesh spacing in one time step. Since the latter requirement depends on the speed of material motions, whereas for an explicit scheme the Courant criterion for this physical system would depend on the sound speed, one can see that the time step can be larger than for an explicit scheme by the ratio of the sound speed to the fluid speed (i.e. the inverse of the Mach number). This ratio is very large for the radiative interior of the Sun. The viscosity must also be large enough to ensure stability, which will be the case provided $\nu \Delta t / \Delta z^2 \geq 1$ where Δz is the mesh spacing.

6.6.3 Codes for computing astrophysical flows

Writing a computer program to solve complex astrophysical fluid-flow problems is a major undertaking. If there is already a code that is suitable for solving the problem, it may be preferable to use that. There are a number of codes that solve time-dependent astrophysical fluid-flow problems, including MHD effects: these may be found using the WWW. They include the codes ZEUS (Stone & Norman 1992), FLASH (Fryxell *et al.* 2000) and NIRVANA (Ziegler 1998) and the PENCIL code (Brandenburg & Dobler 2002). The ZEUS code provides a good starting point for many astrophysical fluid dynamical problems: it is available for solving problems in both 2D and 3D, and several other codes can trace their heritage back to it.

All these codes solve the continuity and momentum equations (1.4), (1.7), Poisson's equation (1.15) in the case of a self-gravitating fluid and, if MHD is included, the MHD induction equation (5.7). Typically they can either use a simple equation of state to relate p and ρ (isothermal conditions or a polytropic relation) or can solve a general energy equation such as Eq. (1.18), including magnetic and viscous dissipation terms as appropriate, for internal energy U, in which case pressure is commonly given by $p = (\gamma - 1)U$. Most codes can incorporate both physical and artificial viscosity; some can include Coriolis and centrifugal forces to represent the momentum equation in a rotating frame.

A quite common approach used by ZEUS-type codes is the adoption of a staggered-mesh formalism in which fluid variables are defined at different locations within a numerical cell (cf. Section 6.6.1). Scalar quantities (e.g. p and ρ) are cell-centred and vector quantities (e.g. u) are face-centred. Pseudo-vectors such as electric field E are edge-centred, while diagonal and off-diagonal elements of rank-2 tensors (e.g. stress) are respectively cell- and face-centred.

Most state-of-the art codes utilise operator splitting. Several of the codes use a second-order flux-conservative finite-volume scheme for the advection part of the equations. NIRVANA also for example uses the ADI method for the heat conduction part of equations and to solve Poisson's equation: a second-order spatial discretization is used for terms in the momentum and energy equations, and source terms in these equations are advanced using an explicit Euler scheme.

The method of characteristics is used by some codes e.g. in solving the induction equation, and adaptive mesh refinement is used to give finer grid resolution where required, based on a nested grid technique (Berger & Oliger 1984). Adaptive mesh techniques are very important in state-of-the-art fluid codes, but space does not permit us to go into these details further here.

For further discussion of the higher-order schemes used in the PENCIL code, and considerations of numerical simulations of MHD and turbulence, see Brandenburg (2003).

6.7 Smoothed Particle Hydrodynamics

Smoothed particle hydrodynamics (SPH) is a very different approach from finite differences, but it has been applied to a wide variety of astrophysical problems. SPH is a particle method, the fluid being represented by a finite set of particles which move as material fluid elements according to the equations of fluid dynamics. The continuum properties of the fluid are recovered using summation interpolants over the set of particles and these continuum properties are used to calculate the forces that govern the motion of the particles. The equation of motion and the energy equation become ordinary differential equations for the time evolution of each particle. SPH does not need a grid to calculate spatial derivatives; but it can be combined with grid-based representations of e.g. magnetic fields.

Central to the SPH representation of continuum fluid properties as summation interpolants over the particles is the interpolating kernel $W(\boldsymbol{r}; h)$. Many choices are possible for the function W. Here h is a parameter defining roughly the spatial extent over which fluid properties are smoothed over the set of particles. Kernels W have the properties that

$$
\left.
\begin{array}{rcl}
\int W(\boldsymbol{r}; h)\mathrm{d}V & = & 1 \\[2mm]
\lim_{h \to 0} W(\boldsymbol{r}; h) & = & \delta(\boldsymbol{r})
\end{array}
\right\}
\tag{6.48}
$$

where $\delta(\boldsymbol{r})$ is the Dirac delta function in three dimensions, and in the integral $\boldsymbol{r}$ ranges over all space. One choice of W with these properties is the Gaussian, $W(\boldsymbol{r}; h) = (h\sqrt{\pi})^{-3}e^{-r^2/h^2}$, where $r = |\boldsymbol{r}|$. This has the disadvantage, however, that formally the kernel extends infinitely far from position r. A kernel that has "compact support", i.e. it is zero beyond a certain distance away from r, is that proposed by Monaghan & Lattanzio (1985) based on spline functions:

$$W(r;h) = \frac{\sigma}{h^d} \begin{cases} 1 - \frac{3}{2}\frac{r^2}{h^2} + \frac{3}{4}\frac{r^4}{h^4} & \text{if } 0 \le r \le h, \\ \frac{1}{4}\left(2 - \frac{r}{h}\right)^3 & \text{if } h \le r \le 2h, \\ 0 & \text{otherwise,} \end{cases} \quad (6.49)$$

where d is the number of spatial dimensions, and normalization constant σ is equal to $2/3$ if $d = 1$, $10/7\pi$ if $d = 2$ and $1/\pi$ if $d = 3$.

Consider then representing the fluid with a set of particles of mass m_j at positions r_j. (Throughout this section, subscripts i and j will be used to label particles, not to represent vector components.) The density is estimated everywhere as

$$\rho(r) = \sum_j m_j W(r - r_j; h) . \quad (6.50)$$

For a general property $f(r)$ of the fluid, the summation interpolant is

$$f(r) = \sum_j \frac{m_j f_j}{\rho_j} W(r - r_j; h) \quad (6.51)$$

where for any quantity f_j is the value of f at r_j, and can thus be thought of as the value of that quantity associated with the jth particle.

From (6.51), the gradient of f can be estimated as

$$\nabla f(r) = \sum_j \frac{m_j f_j}{\rho_j} \nabla W(r - r_j; h) \quad (6.52)$$

where ∇ denotes the gradient operator with respect to the first argument of W. (It is here assumed that h is independent of position. In fact h can be implemented to vary both in space and time – e.g. $h \propto \rho^{-1/3}$ – but this is beyond our present discussion.) In practice, however, to obtain higher accuracy one instead uses (Monaghan 1992)

$$\rho \nabla f = \nabla(\rho f) - f \nabla \rho$$

to write ∇f e.g. at $r = r_i$ as

$$(\nabla f)_i = \rho_i^{-1} \sum_j m_j (f_j - f_i) \nabla W(r_i - r_j; h) . \tag{6.53}$$

Likewise, $\nabla \cdot v$ is better calculated from $(\nabla \cdot \rho v - v \cdot \nabla \rho)/\rho$ so

$$(\nabla \cdot v)_i = \rho_i^{-1} \sum_j m_j (v_j - v_i) \cdot \nabla W((r_i - r_j; h) . \tag{6.54}$$

Here v_i is the velocity of the ith particle.

The particles' positions and velocities are naturally related by

$$\frac{dr_i}{dt} = v_i \tag{6.55}$$

which is how the particle positions are updated. Practically, the right-hand side may be softened by including an average of velocities over neighbouring particles (Monaghan 1992), but this leads to poor energy conservation.

The velocities of the particles are updated in accordance with the momentum equation, e.g.,

$$\frac{dv_i}{dt} = -\left(\frac{1}{\rho}\nabla p\right)_i + g(r_i) . \tag{6.56}$$

The pressure gradient can be estimated using (6.53) with $f = p$, and this indeed has the property that ∇p vanishes identically when the pressure is uniform. However, it has the disadvantage that momentum is not conserved exactly, and for this reason a symmetrized pressure-gradient term may be preferred (Monaghan 1992). Using $\rho^{-1}\nabla p = \nabla(p/\rho) + \rho^{-2}p\nabla\rho$ one can derive

$$\left(\frac{1}{\rho}\nabla p\right)_i = \sum_j m_j \left(\frac{p_i}{\rho_i^2} + \frac{p_j}{\rho_j^2}\right) \nabla W(r_i - r_j; h) . \tag{6.57}$$

The gravitational acceleration due to self-gravity of the fluid may be written as

$$g(r_i) = \sum_j \frac{G\tilde{m}_j}{|r_j - r_i|^2} \frac{(r_j - r_j)}{|r_j - r_i|} . \tag{6.58}$$

It might seem that one should use the mass m_j of the jth particle in the sum in (6.58) and indeed this is appropriate if particles i and j are far apart. But note that (6.50) implies that the mass density of particle j has distribution $m_j W(r - r_j; h)$ and hence is not pointlike. In particular, if

$W(r; h)$ is spherically symmetric in r then by Newton's sphere theorem the appropriate effective mass $\tilde{m}_j$ to use in (6.58) is

$$\tilde{m}_j = m_j \int_{|r-r_j|<|r_i-r_j|} W(r - r_j; h)\mathrm{d}r \ . \tag{6.59}$$

Expression (6.58) is typically evaluated using a hierarchical N-body tree code: see Hernquist & Katz (1989).

Artificial viscosity may be introduced into (6.56) by adding to $\rho_i^{-2} p_i + \rho_j^{-2} p_j$ in (6.57) a term Π_{ij} (cf. Section 6.6.1) where for example

$$\Pi_{ij} = \begin{cases} (-\alpha \bar{c}_{ij} \mu_{ij} + \beta \mu_{ij}^2)/\bar{\rho}_{ij} & \text{if } v_{ij} \cdot r_{ij} < 0 \\ 0 & \text{otherwise} \end{cases} \tag{6.60}$$

and

$$\mu_{ij} = (h v_{ij} \cdot r_{ij})/(r_{ij}^2 + \eta^2) \ . \tag{6.61}$$

Here $\bar{\rho}_{ij}$ and $\bar{c}_{ij}$ are the averages of density and sound speed at particles i and j, and $r_{ij} = r_i - r_j$, $v_{ij} = v_i - v_j$. Monaghan (1992) recommends parameter choices $\alpha = 1$, $\beta = 2$, $\eta = 0.1h$. The viscosity only comes into effect if a pair of particles are approaching one another ($v_{ij} \cdot r_{ij} < 0$), and it is only significant when the particles have a separation of $O(\eta)$ or less. The viscosity term is discussed further in Monaghan's review.

The internal energy equation e.g. for the adiabatic case

$$\frac{\mathrm{d}U_i}{\mathrm{d}t} = \left(\frac{-p}{\rho}\nabla \cdot v\right)_i \tag{6.62}$$

can be estimated as

$$\frac{\mathrm{d}U_i}{\mathrm{d}t} = \frac{p_i}{\rho_i^2}\sum_j m_j(v_i - v_j) \cdot \nabla W(r_i - r_j; h) \tag{6.63}$$

(see Benz 1990). Source terms for heating or heat loss can be introduced. Various equations of state may be also introduced straightforwardly.

The continuity equation need not be implemented, as the evolution of density is given implicitly by (6.50) as

$$\rho_i = \sum_j m_j W(r_i - r_j; h) \tag{6.64}$$

though Monaghan (1992) does argue in favour of implementing it explicitly.

Because SPH reduces the evolution equations to ordinary differential equations, any of various standard numerical solution techniques (e.g.

leapfrog or predictor-corrector) can be used to integrate the system forward in time with appropriate conditions on the timestep used (e.g. Monaghan 1989).

The advantage of SPH is its flexibility in application to complex problems and geometries, and it has an attractive intuitive feel. It has been applied, for instance, to modelling collisions of stars, cloud fragmentation and collapse during star formation, accretion disks and jets.

Errors are not so straightforward to estimate theoretically with SPH as with finite-difference schemes, however. The errors should decrease as the number N of particles increases, though not generally as fast as N^{-1}. An error estimate has been derived by Niederreiter (1978). Arguments suggest that the errors decrease at a rate somewhere between $N^{-\frac{1}{2}}$ and N^{-1}, as the distribution of particles is not random but is often "lattice-like" (R. P. Nelson, private communication).

Chapter 7

Planetary Atmosphere Dynamics

The study of the dynamics of the Earth's atmosphere and oceans is the domain of *geophysical fluid dynamics*. Because of its importance for understanding and predicting man's environment, as well as for its own intrinsic interest, geophysical fluid dynamics is a well developed subject. The considerations of geophysical fluid dynamics are of course applicable to other planets; and because of the advanced development of the field it may also have much to offer the study of solar and stellar interior dynamics.

In this chapter we can only begin to touch a few key considerations of geophysical fluid dynamics. An excellent introduction to the subject is the book by Pedlosky (1979); other good books include those by Cushman-Roisin (1994) at an introductory level and by Gill (1982) at a more advanced level.

7.1 The Importance of Rotation: the Rossby Number

Geophysical fluid dynamics is mostly concerned with large-scale motions in the atmosphere and oceans, and we similarly are primarily concerned here with large-scale motions. By large-scale, we mean motions for which the effects of the planet's rotation are important.

A measure of the importance of rotation on the dynamics of the fluid is the *Rossby number*. Consider motions that have typical speed U and take place on a characteristic length scale L. For example, L might be the size of a convection cell or the typical distance between large-scale pressure peaks and troughs in the atmosphere. The time taken for the fluid to travel distance L at speed U is L/U; we may anticipate that the effects of rotation will be unimportant if that timescale is much smaller than the rotation period of the planet. Put the other way, we expect rotation to be important if $L/U \gtrsim 1/\Omega$, where Ω is the angular speed of the planet's rotation. We may define the non-dimensional Rossby number as the ratio of these two timescales:

$$\text{Ro} = \frac{U}{L\Omega} \, . \tag{7.1}$$

Rotation is important if $\text{Ro} \lesssim 1$.

Large-scale flows are those for which L is sufficiently large that Ro is of order unity or smaller. Indeed the importance of the Coriolis force on large-scale flows is a characteristic of geophysical fluid dynamics.

Since the rotation plays an important role, it is appropriate to work in a frame rotating with the surface of the planet. The motions in the planetary atmosphere, like the winds and currents we are familiar with on Earth, are perturbations to the solid-body rotation that corresponds to $u = 0$ in the rotating frame. From Eqs. (3.3) and (3.8), we may write the momentum equation in that frame as

$$\frac{\partial u}{\partial t} + u \cdot \nabla u + 2\Omega \times u = -\frac{1}{\rho}\nabla p - \nabla \Phi \tag{7.2}$$

where Φ is the effective gravitational potential, including the centrifugal term, and $\partial/\partial t$ is the time derivative at constant position fixed in the rotating frame. Henceforth in this chapter we interpret all quantities as being measured in the rotating frame unless stated otherwise. Viscosity has here been assumed negligible: we return to the role of viscosity in Section 7.8.

7.2 Relative and Absolute Vorticity

A key quantity for understanding the dynamics when rotation is important is the vorticity

$$\omega = \nabla \times u \tag{7.3}$$

(see Section 1.9). We call the vorticity (7.3) measured in the frame rotating with angular velocity Ω the *relative vorticity*. This is related to the vorticity ω_a measured in an inertial frame, which we call the *absolute vorticity*, by

$$\omega_a = 2\Omega + \omega \, . \tag{7.4}$$

The momentum equation (7.2) may be rewritten

$$\frac{\partial \boldsymbol{u}}{\partial t} = \boldsymbol{u} \times \boldsymbol{\omega}_a - \frac{1}{\rho} \nabla p - \nabla \Phi . \tag{7.5}$$

Taking the curl of (7.5) gives the analogues of Eqs. (1.54) and (1.55) (with $\nabla \times \boldsymbol{f} = 0$):

$$\frac{\partial \boldsymbol{\omega}}{\partial t} = \nabla \times (\boldsymbol{u} \times \boldsymbol{\omega}_a) + \frac{1}{\rho^2} \nabla \rho \times \nabla p \tag{7.6}$$

and

$$\frac{\partial \boldsymbol{\omega}}{\partial t} + \boldsymbol{u} \cdot \nabla \boldsymbol{\omega} = (\boldsymbol{\omega}_a \cdot \nabla) \boldsymbol{u} - (\nabla \cdot \boldsymbol{u}) \boldsymbol{\omega}_a + \frac{1}{\rho^2} \nabla \rho \times \nabla p \tag{7.7}$$

which we can refer to as the vorticity equation in a rotating frame. Since $2\boldsymbol{\Omega}$ is a constant, $\boldsymbol{\omega}$ can also be replaced by $\boldsymbol{\omega}_a$ in the left-hand sides of (7.6) and (7.7) – indeed, substituting $\boldsymbol{\omega}_a$ from (7.4) into Eqs. (1.54) and (1.54) would be an alternative derivation of (7.6) and (7.7). Thus for example

$$\frac{\partial \boldsymbol{\omega}_a}{\partial t} + \boldsymbol{u} \cdot \nabla \boldsymbol{\omega}_a = (\boldsymbol{\omega}_a \cdot \nabla) \boldsymbol{u} - (\nabla \cdot \boldsymbol{u}) \boldsymbol{\omega}_a + \frac{1}{\rho^2} \nabla \rho \times \nabla p . \tag{7.8}$$

In the rotating frame we may define the circulation or vortex tube strength, as in Section 1.9, as $\oint_C \boldsymbol{u} \cdot \mathrm{d}\boldsymbol{r} \equiv \int \boldsymbol{\omega} \cdot \boldsymbol{n} \mathrm{d}S$.

It is straightforward to generalize the derivation (1.58) of $\frac{\mathrm{d}}{\mathrm{d}t} \oint_C \boldsymbol{u} \cdot \mathrm{d}\boldsymbol{r}$ for the rotating case, and show that the right-hand side of Eq. (1.58) gains an additional term

$$\oint_C (-2\boldsymbol{\Omega} \times \boldsymbol{u}) \cdot \mathrm{d}\boldsymbol{r} . \tag{7.9}$$

Thus there is an additional way for the circulation around C (equivalently, the flux of vorticity through C) to be changed in a rotating system, namely due to the deflection of the flow by Coriolis force. If $\boldsymbol{\Omega}$ is positive through C (defined as usual in an anticlockwise sense), and $\boldsymbol{u}$ is outwards across C, then the Coriolis term tends to decrease the circulation, tending to cause a flow clockwise around C. Contrariwise, if $\boldsymbol{u}$ is inwards across C then this term tends to increase the circulation with a flow anticlockwise around C. (See the discussion of geostrophic flows in Section 7.5.) If the conditions for Kelvin's circulation theorem hold – i.e., barotropic fluid with no viscosity – the absolute vorticity is preserved but, because of the Coriolis term, the circulation $\oint_C \boldsymbol{u} \cdot \mathrm{d}\boldsymbol{r}$ is not.

7.3   Potential Vorticity

The potential vorticity is a strangely named quantity since it may not even have the dimensions of vorticity. Nonetheless the term is widely used in geophysical fluid dynamics, and its conservation property is very useful.

Let Q be some (as yet unspecified) quantity of the fluid flow and consider the variation of $\rho^{-1}\boldsymbol{\omega}_a \cdot \nabla Q$, where $\boldsymbol{\omega}_a$ is the absolute vorticity. For an inviscid fluid

$$
\frac{D}{Dt}\left(\frac{\boldsymbol{\omega}_a}{\rho} \cdot \nabla Q\right) \equiv \frac{D}{Dt}\left(\frac{\boldsymbol{\omega}_a}{\rho}\right) \cdot \nabla Q + \frac{\boldsymbol{\omega}_a}{\rho} \cdot \frac{D}{Dt}(\nabla Q)
$$

$$
\equiv \frac{D}{Dt}\left(\frac{\boldsymbol{\omega}_a}{\rho}\right) \cdot \nabla Q + \left(\frac{\boldsymbol{\omega}_a}{\rho} \cdot \nabla\right)\frac{DQ}{Dt} - \left(\frac{\boldsymbol{\omega}_a}{\rho} \cdot \nabla\boldsymbol{u}\right) \cdot \nabla Q
$$

$$
= \frac{1}{\rho^3}\left(\nabla\rho \times \nabla p\right) \cdot \nabla Q + \left(\frac{\boldsymbol{\omega}_a}{\rho} \cdot \nabla\right)\frac{DQ}{Dt} \tag{7.10}
$$

using Eq. (7.8). Now, if $DQ/Dt = 0$, i.e. Q is a conserved quantity, and if $(\nabla\rho \times \nabla p) \cdot \nabla Q = 0$, then the *potential vorticity*

$$
\Pi = \frac{\boldsymbol{\omega}_a}{\rho} \cdot \nabla Q \tag{7.11}
$$

is conserved, i.e.,

$$
\frac{D\Pi}{Dt} = 0 . \tag{7.12}
$$

Note that the condition $(\nabla\rho \times \nabla p) \cdot \nabla Q = 0$ is satisfied if the fluid is barotropic (so $\nabla\rho \times \nabla p = 0$) or if Q is a function of p and ρ only. (It is a simple exercise to show that $Q = Q(p, \rho)$ implies that $(\nabla\rho \times \nabla p) \cdot \nabla Q = 0$.)

An implication of the conservation of potential vorticity is that if surfaces of constant Q move further apart, so $|\nabla Q|$ decreases, the component of $\boldsymbol{\omega}_a/\rho$ in the direction of ∇Q must increase in magnitude to keep Π constant for a fluid element.

7.4   Baroclinicity and the Thermal Wind Equation

We have not yet used the fact that the Rossby number is small. Let U and L respectively be characteristic speed and length-scale of the flow. We assume that the relative vorticity $\boldsymbol{\omega}$ is of order U/L. If the Rossby number is very small then $|\boldsymbol{\omega}| \ll |\boldsymbol{\Omega}|$ and so substituting $\boldsymbol{\omega}_a \simeq 2\boldsymbol{\Omega}$ in the vorticity equation (7.7) gives the approximation

$$\frac{\partial \omega}{\partial t} + \boldsymbol{u} \cdot \nabla \omega = 2\boldsymbol{\Omega} \cdot \nabla \boldsymbol{u} - 2\boldsymbol{\Omega}(\nabla \cdot \boldsymbol{u}) + \frac{\nabla \rho \times \nabla p}{\rho^2} . \qquad (7.13)$$

If the rotation period is much smaller than any other timescale of the flow (or to put it more formally, if the temporal Rossby number $\mathrm{Ro}_T \equiv (\Omega T)^{-1} \ll 1$, where T is the local timescale for the flow to change), the two terms on the left-hand side of (7.13) are negligible compared with the sum of terms involving $\boldsymbol{\Omega}$, and so (7.13) is further approximated as

$$2\boldsymbol{\Omega} \cdot \nabla \boldsymbol{u} - 2\boldsymbol{\Omega}(\nabla \cdot \boldsymbol{u}) = -\frac{\nabla \rho \times \nabla p}{\rho^2} . \qquad (7.14)$$

In other words, there is a balance between the source terms on the right-hand side of vorticity equation (7.13) which might otherwise cause the relative vorticity to grow to violate the assumption of small Rossby number.

Let the direction along $\boldsymbol{\Omega}$ be the z-direction; and let $\nabla_\perp$ and $\boldsymbol{u}_\perp$ be the components of ∇ and $\boldsymbol{u}$ perpendicular to $\boldsymbol{\Omega}$. Then (7.14) can be split into its components perpendicular to and parallel to $\boldsymbol{\Omega}$ respectively as:

$$2\boldsymbol{\Omega} \cdot \nabla \boldsymbol{u}_\perp = -\frac{1}{\rho^2} \left\{ \frac{\partial p}{\partial z} (\nabla_\perp \rho) \times \boldsymbol{e}_z - \frac{\partial \rho}{\partial z} (\nabla_\perp p) \times \boldsymbol{e}_z \right\} , \qquad (7.15)$$

$$-2\boldsymbol{\Omega}(\nabla \cdot \boldsymbol{u}_\perp) = -\frac{1}{\rho^2}(\nabla_\perp \rho) \times (\nabla_\perp p) . \qquad (7.16)$$

Equation (7.15) relates the variations along the rotation axis of the velocity perpendicular to $\boldsymbol{\Omega}$ to density variations. These density variations are frequently connected with thermal variations; hence (7.15) is called the *thermal wind equation*.

We note the particular case, which we have already seen when considering the rotation of the solar interior (Section 3.6), when $\boldsymbol{u} = u_\phi \boldsymbol{e}_\phi$ is in the azimuthal direction about the $\boldsymbol{\Omega}$ axis, and u_ϕ, p and ρ are independent of azimuthal coordinate ϕ. Then $\nabla \cdot \boldsymbol{u} = 0$ and Eq. (7.14) yields

$$(2\boldsymbol{\Omega} \cdot \nabla u_\phi) \boldsymbol{e}_\phi = -\frac{\nabla \rho \times \nabla p}{\rho^2} . \qquad (7.17)$$

In the case that the fluid is barotropic, so $\nabla \rho \times \nabla p = 0$, the right-hand side of (7.17) vanishes and so $\partial u_\phi / \partial z = 0$, i.e. u_ϕ does not vary in the z-direction (i.e. parallel to the rotation axis). This is one statement of the *Taylor-Proudman theorem*.

An alternative statement is that if the fluid is barotropic and the flow is incompressible ($\nabla \cdot \boldsymbol{u} = 0$) then from Eq. (7.14)

$$2\boldsymbol{\Omega} \cdot \nabla \boldsymbol{u} = 0, \tag{7.18}$$

i.e. there is no variation in $\boldsymbol{u}$ in the direction parallel to the rotation axis. This too is known as the *Taylor-Proudman theorem*. The additional assumption of incompressibility essentially just imposes that $\partial u_z / \partial z$ is zero, since (7.16) already implies that the horizontal flow has zero divergence for a barotropic flow.

7.5 Geostrophic Motion

We have seen in the previous section that at small Rossby number the vorticity equation can be approximated to Eq. (7.13) and further approximated to Eq. (7.14), from which follow the thermal-wind equation and the Taylor-Proudman theorem. By similar reasoning, the momentum equation (7.2) can also usefully be approximated for small Rossby number. Assuming that the relative acceleration and nonlinear acceleration terms on the left of Eq. (7.2) are small compared with the Coriolis acceleration $2\boldsymbol{\Omega} \times \boldsymbol{u}$, the momentum equation reduces to

$$2\boldsymbol{\Omega} \times \boldsymbol{u} = -\frac{1}{\rho}\nabla p - \nabla \Phi. \tag{7.19}$$

Of course, Eq. (7.14) is derivable directly from this equation by taking its curl.

Let us resolve (7.19) into components in a spherical polar coordinate system (r, θ, ϕ), where the polar axis is the rotation axis. The r-, θ- and ϕ-components of (7.19) are respectively

$$-2\Omega \sin\theta\, u_\phi = -\frac{1}{\rho}\frac{\partial p}{\partial r} - g \tag{7.20}$$

$$-2\Omega \cos\theta\, u_\phi = -\frac{1}{\rho r}\frac{\partial p}{\partial \theta} \tag{7.21}$$

$$2\Omega \left(\cos\theta\, u_\theta + \sin\theta\, u_r\right) = -\frac{1}{\rho r \sin\theta}\frac{\partial p}{\partial \phi}. \tag{7.22}$$

Note that $\nabla\Phi$ has been taken to be $g\boldsymbol{e}_r$, i.e. neglecting any deviation of the effective gravitational acceleration from the radial direction. Note also that here θ is the standard polar angle θ, i.e. the co-latitude, whereas commonly in the geophysical literature θ is defined to be the latitude: thus in some geophysical texts equations like the above appear but with the roles of $\cos\theta$ and $\sin\theta$ reversed and with a sign change in the derivative with respect to θ.

A typical application is to the atmosphere of the Earth (or other terrestrial planet). The fluid region comprises a thin layer overlying the surface of the planet and it is convenient to introduce local Cartesian coordinates (x, y, z), a right-handed set where x increases eastwards, y increases northwards and z increases upwards. With respect to these coordinates the velocity is (u, v, w). In terms of u_r, u_θ, u_ϕ, we have $u = u_\phi$, $v = -u_\theta$, $w = u_r$. Let the thickness of the fluid layer be H, the characteristic horizontal velocity be U and the characteristic horizontal scale on which velocity and pressure vary be L. Since we are interested in large-scale motion, we have $H/L \ll 1$. The fact that the layer remains thin and does not spread places a constraint the large-scale fluid trajectories and hence on the magnitude of the vertical velocity component w:

$$w = O\left(\frac{H}{L}U\right) \ll (u^2 + v^2)^{1/2} \ .$$

Hence the motion is essentially purely horizontal. The term in (7.22) involving the vertical (radial) velocity can therefore be neglected. A second simplification comes from considering orders of magnitude of terms in Eqs. (7.20) and (7.21). Let p', ρ' be the departures of p and ρ from the values they would take in the absence of fluid motions. Then (7.22) implies that $p'/\rho L \sim \Omega U$. The term on the left of (7.20) is also of this order, but if the right-hand side of (7.20) does not balance itself exactly then it is of order $p'/\rho H = (\Omega U)(L/H) \gg \Omega U$. The conclusion is that the term on the left of (7.20) cannot contribute significantly to the balance of the equation since it is too small by a factor $H/L \ll 1$, and so it is negligible. Hence Eqs. (7.20)–(7.22) simplify to

$$-\rho g = \frac{\partial p}{\partial r} \tag{7.23}$$

$$fu = \frac{1}{\rho r}\frac{\partial p}{\partial \theta} \tag{7.24}$$

$$fv = \frac{1}{\rho r \sin\theta}\frac{\partial p}{\partial \phi} \tag{7.25}$$

where we have introduced the *Coriolis parameter* f defined by

$$f = 2\Omega\cos\theta \ . \tag{7.26}$$

In the local Cartesian coordinates, Eqs. (7.23)–(7.25) are written

$$-\rho g = \frac{\partial p}{\partial z} \tag{7.27}$$

$$fu = -\frac{1}{\rho}\frac{\partial p}{\partial y} \tag{7.28}$$

$$fv = \frac{1}{\rho}\frac{\partial p}{\partial x}. \tag{7.29}$$

Before proceeding further we note that, since p', ρ' are the departures from hydrostatic equilibrium, Eq. (7.27) implies that

$$\rho' = \mathrm{O}\left(p'/gH\right) = \mathrm{O}\left(\rho\Omega U L/gH\right), \tag{7.30}$$

so

$$\rho'/\rho = \mathrm{O}\left(\mathrm{Ro} \times \Omega^2 L^2/gH\right). \tag{7.31}$$

Provided that $\Omega^2 L^2/gH$ is not bigger than order unity, for small Rossby number the relative density perturbations must therefore be small. It follows that perturbations to ρ may then be neglected in Eqs. (7.24), (7.25) and Eqs. (7.28), (7.29).

The balance between horizontal velocities and horizontal pressure gradients in Eqs. (7.24), (7.25) or Eqs. (7.28), (7.29) is called the *geostrophic approximation*. In vector form, it gives the horizontal component u_h of the velocity as

$$u_h = \frac{1}{fp}e_z\times\nabla p. \tag{7.32}$$

Such a flow, determined by a balance between Coriolis forces and pressure gradients, is called *geostrophic*.

The remarkable conclusion from Eq. (7.32) is that, unlike in a non-rotating fluid where the flow would normally be in the direction of the pressure force, in a geostrophic flow the horizontal velocity is perpendicular to ∇p; thus the flow is along lines of constant pressure, i.e. along isobars. Thus all geostrophic flows are isobaric. This pattern can commonly be seen, at least to a good approximation, on meteorological maps. In the northern hemisphere, f is positive and (7.32) implies that about a region where pressure is high (an anticyclone) the flow is clockwise, whereas about a low-pressure area (a depression or cyclone) the flow is anticlockwise. In the southern hemisphere the Coriolis parameter f is negative and so the flow directions are reversed. The sense of direction of the motion can be simply recalled by considering a fluid parcel attempting to flow away from a region of high pressure, or towards a low-pressure region, and thinking in what direction the Coriolis force would act.

The geostrophic flow can also be expressed in terms of the slope of surfaces of constant pressure. First we recall a rule for taking ratios of partial derivatives. Since

$$dp = \left(\frac{\partial p}{\partial x}\right)_{y,z} dx + \left(\frac{\partial p}{\partial y}\right)_{z,x} dy + \left(\frac{\partial p}{\partial z}\right)_{x,y} dz$$

can be re-arranged to give

$$dz = \left(\frac{\partial p}{\partial z}\right)_{x,y}^{-1} dp - \frac{(\partial p/\partial x)_{y,z}}{(\partial p/\partial z)_{x,y}} dx - \frac{(\partial p/\partial y)_{z,x}}{(\partial p/\partial z)_{x,y}} dy$$

it follows for instance that

$$-\frac{(\partial p/\partial x)_{y,z}}{(\partial p/\partial z)_{x,y}} \equiv \left(\frac{\partial z}{\partial x}\right)_{y,p} , \tag{7.33}$$

which is a straightforward extension of identity (1.26). Now dividing Eq. (7.29) by (7.27) gives

$$fv = -g\frac{\partial p/\partial x}{\partial p/\partial z} = g\left(\frac{\partial z}{\partial x}\right)_{y,p} . \tag{7.34}$$

Similarly, dividing Eq. (7.28) by (7.27),

$$fu = -g\left(\frac{\partial z}{\partial y}\right)_{x,p} . \tag{7.35}$$

These can be combined in a single vector equation

$$\boldsymbol{u}_h = \frac{g}{f}\boldsymbol{e}_z \times (\nabla z)_p , \tag{7.36}$$

i.e. the geostrophic velocity is given by the horizontal derivatives of the height of the surfaces of constant pressure.

The Coriolis parameter varies with latitude only, so in e.g. Eq. (7.29) it is a function of y but is independent of x and z. Differentiating (7.29) with respect to z, and using (7.27) twice, yields

$$f\frac{\partial v}{\partial z} = -\frac{g}{\rho}\frac{\partial \rho}{\partial x} + \frac{g}{\rho}\frac{\partial \rho}{\partial z}\frac{\partial p}{\partial x}\bigg/\frac{\partial p}{\partial z} ; \tag{7.37}$$

whence, using identity (1.26) on the last term,

$$f\frac{\partial v}{\partial z} = -\frac{g}{\rho}\left(\frac{\partial\rho}{\partial x}\right)_{y,p}.$$ (7.38)

Likewise, differentiating Eq. (7.28) with respect to z yields

$$f\frac{\partial u}{\partial z} = \frac{g}{\rho}\left(\frac{\partial\rho}{\partial y}\right)_{x,p}.$$ (7.39)

These two equations can be expressed in vector form as

$$\frac{\partial\boldsymbol{u}_h}{\partial z} = -\frac{g}{f}\boldsymbol{e}_z\times\frac{1}{\rho}\left(\nabla\rho\right)_p.$$ (7.40)

Hence the variation of horizontal velocity with z is determined by the horizontal variation of density on constant-pressure surfaces. A corollary of Eq. (7.40) is that if density is a function only of pressure, so $(\nabla\rho)_p = 0$, then $\boldsymbol{u}_h$ does not vary with height. Hence geostrophy implies the Taylor-Proudman theorem.

It may also be noted that for a perfect gas

$$\frac{1}{\rho}\left(\nabla\rho\right)_p = \frac{1}{T}\left(\nabla T\right)_p;$$

hence $\partial\boldsymbol{u}_h/\partial z$ may also be expressed in terms of horizontal derivatives of temperature on constant-pressure surfaces.

A final comment is that although the geostrophic approximation provides an extremely useful diagnostic tool, it describes the flow with a lower-order system of equations than the full fluid equations. Hence it is not possible in general to satisfy the full set of boundary conditions imposed on the flow. Thus the flow cannot generally be geostrophic everywhere. It breaks down where the horizontal length scale L is not large as assumed. For example, the gulf stream is *ageostrophic*, as it must be thin (cf. also *fronts* in meteorology).

7.6   Some Approximate Models

Since the full equations of geophysical fluid dynamics are so complicated, various systems of model equations that approximate the dynamics can be found in the literature. Here we mention just three. As well as modelling atmospheres and oceans, they have also found applicability in modelling a small region inside the Sun.

7.6.1 The shallow-water model

We relax the assumption in Section 7.5 that the Rossby number is small, but introduce the additional assumption that the fluid is homogeneous. We again work in a local Cartesian coordinate system. The application in mind is the situation illustrated in Fig. 7.1: a fluid layer with free upper surface (so pressure constant there, say p_a) flows over a rigid base whose height varies with x and y as $h_B(x, y)$; the thickness of the fluid layer is $h(x, y, t)$. We anticipate that the horizontal velocity components are independent of vertical coordinate z, an assumption that we justify after deriving Eq. (7.45) for the pressure. The Eqs. (7.27)–(7.29) are replaced by

$$\frac{\partial u}{\partial t} + u\frac{\partial u}{\partial x} + v\frac{\partial u}{\partial y} - fv = -\frac{1}{\rho_0}\frac{\partial p}{\partial x}. \tag{7.41}$$

$$\frac{\partial v}{\partial t} + u\frac{\partial v}{\partial x} + v\frac{\partial v}{\partial y} + fu = -\frac{1}{\rho_0}\frac{\partial p}{\partial y} \tag{7.42}$$

$$g = -\frac{1}{\rho_0}\frac{\partial p}{\partial z} \tag{7.43}$$

where ρ_0 is the homogeneous density, and the continuity equation is

$$\frac{\partial u}{\partial x} + \frac{\partial v}{\partial y} + \frac{\partial w}{\partial z} = 0. \tag{7.44}$$

Equation (7.43) can be integrated immediately with respect to z to give

$$p = -\rho_0 g z + p'$$

where $p'(x, y, t)$ is a "constant of integration". But we know that the surface pressure is p_a, i.e. $p = p_a$ at $z = h + h_B$; this determines the function p' and so

$$p = -\rho_0 g z + \rho_0 g (h + h_B). \tag{7.45}$$

This determines the pressure in terms of h (and h_B). Note that (7.45) implies that $\partial p/\partial x$ and $\partial p/\partial y$ are independent of z. Hence the horizontal acceleration due to pressure has no z-dependence. Thus if u and v are independent of z initially, as we shall suppose, then they remain independent of z. This justifies our dropping $w\partial u/\partial z$ (and similarly for v) when writing the left-hand sides of Eqs. (7.41), (7.42).

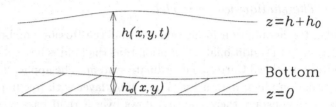

Fig. 7.1   Set-up for shallow-water model.

The vertical velocity w can be found in terms of h by integrating Eq. (7.44) from $z = h_B$ to $z = h$, noting that u and v are independent of z:

$$\left(\frac{\partial u}{\partial x} + \frac{\partial v}{\partial y}\right) h + w_{z=h+h_B} - w_{z=h_B} = 0 . \qquad (7.46)$$

The trajectory of a fluid element at the surface (top or bottom) must follow that surface; hence

$$w_{z=h_B} = u\frac{\partial h_B}{\partial x} + v\frac{\partial h_B}{\partial y} ,$$

$$w_{z=h+h_B} = \frac{\partial h}{\partial t} + u\frac{\partial}{\partial x}(h + h_B) + v\frac{\partial}{\partial y}(h + h_B) .$$

Substituting these expressions into Eq. (7.46) finally gives

$$\frac{\partial h}{\partial t} + \frac{\partial}{\partial x}(uh) + \frac{\partial}{\partial y}(vh) = 0 , \qquad (7.47)$$

or equivalently

$$\frac{Dh}{Dt} + h\,\nabla\cdot\boldsymbol{u}_h = 0 , \qquad (7.48)$$

where $\boldsymbol{u}_h$ is the horizontal velocity $(u, v, 0)$.

If we consider a fluid column of horizontal cross-sectional area A, then $\nabla \cdot \boldsymbol{u}_h$ is precisely $A^{-1}DA/Dt$. Thus (7.48) states that

$$\frac{1}{h}\frac{Dh}{Dt} + \frac{1}{A}\frac{DA}{Dt} = 0 \,, \tag{7.49}$$

i.e. the volume hA is conserved.

Finally, substituting for pressure p from (7.45) into Eqs. (7.41), (7.42), together with (7.47), gives the shallow-water equations

$$\frac{\partial u}{\partial t} + u\frac{\partial u}{\partial x} + v\frac{\partial u}{\partial y} - fv = -g\frac{\partial}{\partial x}(h + h_B) \,, \tag{7.50}$$

$$\frac{\partial v}{\partial t} + u\frac{\partial v}{\partial x} + v\frac{\partial v}{\partial y} + fu = -g\frac{\partial}{\partial y}(h + h_B) \,, \tag{7.51}$$

$$\frac{\partial h}{\partial t} + \frac{\partial}{\partial x}(uh) + \frac{\partial}{\partial y}(vh) = 0 \,. \tag{7.52}$$

These are used to determine $u(x,y,t)$, $v(x,y,t)$ and $h(x,y,t)$.

7.6.2 *f-plane and β-plane models*

We have seen that the Coriolis parameter f enters into the various formulations of the equations of motion in local Cartesian coordinates. From its definition (7.26) it is evident that f depends on latitude. One sees in the literature reference to two families of approximate models. In *f-plane* models, the Coriolis parameter is treated as a constant, f_0 say. Its constant value is chosen to be representative of the typical co-latitude θ_0 of the region being modelled.

A next level of approximation can be obtained from a Taylor expansion of f about $\theta = \theta_0$:

$$f = 2\Omega\cos\theta_0 + 2\Omega\sin\theta_0 \,(\theta_0 - \theta) + \mathrm{O}\,(\theta - \theta_0)^2 \,.$$

Retaining only the first two terms one obtains the β-plane approximation: in β-plane models, f is approximated as

$$f = f_0 + \beta_0 y \tag{7.53}$$

with f_0 and β_0 constants. In general it is assumed that $|\beta_0 y| \ll |f_0|$.

7.7   Waves

Waves are an important phenomenon in fluid dynamical systems. We have
already seen some wave motions in Chapter 2. We now consider small-
amplitude motions in geophysical fluid dynamics. For simplicity, we work
in the framework of the shallow-water model (Section 7.6.1) with $h_B = 0$.

Let the depth of the layer be $h_0 + \eta$, where h_0 is the equilibrium depth
and $|\eta| \ll |h_0|$. Under the assumption that the velocities u, v are small, so
all quadratic terms in u, v, η can be neglected, the linearized Eqs. (7.50)–
(7.52) become

$$\frac{\partial u}{\partial t} - fv = -g\frac{\partial \eta}{\partial x} \,, \tag{7.54}$$

$$\frac{\partial v}{\partial t} + fu = -g\frac{\partial \eta}{\partial y} \,, \tag{7.55}$$

$$\frac{\partial h}{\partial t} + \frac{\partial}{\partial x}(h_0 u) + \frac{\partial}{\partial y}(h_0 v) = 0 \,. \tag{7.56}$$

Treating f as a constant, we can obtain a *local dispersion relation* by seeking
a solution in which u, v, η vary with x, y, t as

$$\exp(ik_x x + ik_y y - i\omega t) \tag{7.57}$$

(real part understood). Then Eqs. (7.54)–(7.56) yield the algebraic equations

$$\begin{pmatrix} -i\omega & -f & igk_x \\ f & -i\omega & igk_y \\ ik_x h_0 & ik_y h_0 & -i\omega \end{pmatrix} \begin{pmatrix} u \\ v \\ \eta \end{pmatrix} = \begin{pmatrix} 0 \\ 0 \\ 0 \end{pmatrix} \,. \tag{7.58}$$

For a non-trivial solution (i.e. u, v, η not all zero) the determinant of the
matrix must be zero — otherwise (7.58) could be solved by multiplying by
the matrix inverse. Evaluating the determinant yields

$$i\omega \left(\omega^2 - f^2 - gh_0 k_h^2 \right) = 0 \tag{7.59}$$

where

$$k_h^2 = k_x^2 + k_y^2 \,. $$

One solution is $\omega = 0$ (steady solution) to which we return later. The other
is

$$\omega^2 = gh_0 k_h^2 + f^2 \equiv gh_0(k_x^2 + k_y^2) + f^2 \tag{7.60}$$

which we recognise as the dispersion relation of gravity waves in a shallow layer (Section 2.6) but modified by the addition of the f^2 term caused by the rotation.

To obtain a global dispersion relation it is necessary to impose the boundary conditions appropriate to the flow. We shall not go into any detailed cases here, but simply remark on two cases commonly found in the geophysical literature. If the flow takes place in a channel between rigid walls at $y = 0$ and $y = L$, say, then this permits solution with dependence (7.57) provided k_y takes one of a set of discrete values that permit boundary conditions on both walls to be satisfied. For each such value of k_y the dispersion relation is (7.60). These are *Poincaré waves*. The second case is waves in a semi-infinite domain ($y > 0$ say) with a wall at $y = 0$. This admits solutions with $v = 0$ everywhere. The reduced matrix Eq. (7.58) admits non-trivial solutions for u and η provided determinant

$$\begin{vmatrix} -i\omega & igk_x \\ ik_x h_0 & -i\omega \end{vmatrix} = 0 \, ,$$

i.e.

$$\omega^2 = gh_0 k_x^2 \, . \tag{7.61}$$

How does this relate to the local dispersion relation (7.60)? They are consistent provided $gh_0 k_y^2 + f^2 = 0$, i.e. $k_y = \pm if/\sqrt{gh_0}$. One independent solution corresponds to waves growing exponentially with increasing y, the other to waves decaying exponentially. The latter is the physically acceptable solution: thus these waves propagate parallel to the boundary but their amplitude decreases exponentially away from it. These are called *Kelvin waves*.

So far we have treated f as a constant. On large scales, however, the variations of f with coordinate y must be taken into account. We do so with the β-plane approximation (7.53). As well as making a small modification to the above wave motions, this admits a new class of wave solutions of Eqs. (7.54)–(7.56). This class of solutions has low frequency. Consider then a zeroth-order approximation in which we seek solutions to these equations but treating the $\partial/\partial t$ terms and the βy part of f as negligible. The solutions for u and v,

$$u = -\frac{g}{f_0} \frac{\partial \eta}{\partial y} \, , \qquad v = \frac{g}{f_0} \frac{\partial \eta}{\partial x} \, . \tag{7.62}$$

We recognise this as the same as the geostrophic solution. Substituting these zeroth-order solutions into the previously neglected small terms, Eqs. (7.54), (7.55) become

$$\frac{\partial}{\partial t}\left(-\frac{g}{f_0}\frac{\partial \eta}{\partial y}\right) - f_0 v - \beta_0 y\left(\frac{g}{f_0}\frac{\partial \eta}{\partial x}\right) = -g\frac{\partial \eta}{\partial y}, \qquad (7.63)$$

$$\frac{\partial}{\partial t}\left(\frac{g}{f_0}\frac{\partial \eta}{\partial x}\right) + f_0 u + \beta_0 y\left(-\frac{g}{f_0}\frac{\partial \eta}{\partial y}\right) = -g\frac{\partial \eta}{\partial x}, \qquad (7.64)$$

which can be re-arranged to make v and u the subjects of the formulae. These can then in turn be used to eliminate u and v from Eq. (7.56), yielding

$$\frac{\partial \eta}{\partial t} = \frac{gh_0}{f_0^2}\left(\frac{\partial^2}{\partial x^2} + \frac{\partial^2}{\partial y^2}\right)\frac{\partial \eta}{\partial t} + \beta_0 \frac{gh_0}{f_0^2}\frac{\partial \eta}{\partial x}. \qquad (7.65)$$

Unlike Eqs. (7.63) and (7.64), Eq. (7.65) has constant coefficients and so we can seek solutions with exponential dependence (7.57). After re-arrangement, this yields the dispersion relation

$$\omega = \frac{-\beta_0(gh_0/f_0^2)k_x}{1 + (gh_0/f_0^2)k_h^2}. \qquad (7.66)$$

These are called *Rossby waves* or planetary waves. They correspond to the $\omega = 0$ solutions of Eq. (7.59), but we see that the latitudinal dependence of the Coriolis force gives them a non-zero frequency. It is straightforward to show that the maximum frequency of these waves has magnitude $\frac{1}{2}\beta_0\sqrt{gh_0}/f_0$, and this is attained when $|k_x| = |f_0|/\sqrt{gh_0}$ and $k_y = 0$.

7.8   Ekman Layers

A measure of the importance of viscosity, relative to Coriolis effects, is the Ekman number

$$\text{Ek} = \frac{\nu}{\Omega \mathcal{L}^2}$$

where $\mathcal{L}$ is a length characterizing the depth scale of the motion. Mostly in astrophysical fluids $\text{Ek} \ll 1$ and viscosity is unimportant; but, as in other fluid dynamical applications, near boundaries one finds boundary layers in which the effects of viscosity are important. Such boundaries can be fluid-solid interfaces, such as a planet's surface; or they can be fluid-fluid interfaces where a tangential stress force is applied by one fluid on the other, such as the wind on the surface of the ocean. Roughly speaking the thickness d of such a boundary layer is such that the Ekman number based on length scale d is of order unity:

$$\frac{\nu}{\Omega d^2} \sim 1 \,, \qquad d \sim (\nu/\Omega)^{\frac{1}{2}} \,. \qquad (7.67)$$

Let us suppose that the boundary is located at $z = 0$, with the fluid occupying the semi-infinite region on one side of $z = 0$, and that far from the boundary ($|z| \gg d$) the flow velocity is $\boldsymbol{u} = u_\infty \boldsymbol{e}_x$. Since it plays no direct role we shall neglect gravity. In the boundary layer the viscous force term $\nu \nabla^2 \boldsymbol{u}$ cannot be neglected in the momentum equation. However, the horizontal length scale of variation is much larger than d, so $\partial/\partial x$ and $\partial/\partial y$ are small compared with $\partial/\partial z$; hence we retain only the z-derivatives in $\nabla^2 \boldsymbol{u}$. Thus the equations to solve, with viscous term included, are

$$-fv = -\frac{1}{\rho_0}\frac{\partial p}{\partial x} + \nu \frac{\partial^2 u}{\partial z^2} \,, \qquad (7.68)$$

$$fu = -\frac{1}{\rho_0}\frac{\partial p}{\partial y} + \nu \frac{\partial^2 v}{\partial z^2} \,, \qquad (7.69)$$

$$0 = -\frac{1}{\rho_0}\frac{\partial p}{\partial z} \,. \qquad (7.70)$$

Outside the boundary layer the viscous terms vanish and $u \to u_\infty$, $v \to 0$. Hence, outside the layer, Eq. (7.68) implies

$$\frac{\partial p}{\partial x} = 0 \,, \qquad \frac{\partial p}{\partial y} = -\rho_0 f u_\infty \,. \qquad (7.71)$$

Now Eq. (7.70) implies that the pressure field is independent of z, and therefore so too are $\partial p/\partial x$, $\partial p/\partial y$. Thus expressions (7.71) apply also in the boundary layer; and substituting these into Eqs. (7.68), (7.69) gives:

$$-fv = \nu \frac{\partial^2 u}{\partial z^2} \,, \qquad (7.72)$$

$$f(u - u_\infty) = \nu \frac{\partial^2 v}{\partial z^2} \,. \qquad (7.73)$$

We wish to solve these equations in the boundary layer with boundary conditions that $u \to u_\infty$, $v \to 0$ as $|z| \to \infty$ and appropriate boundary conditions at $z = 0$. Since the equations have constant coefficients, we anticipate independent solutions $u - u_\infty$, v proportional to $\exp(\lambda|z|)$. Then

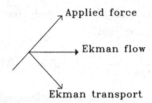

Fig. 7.2   Direction of Ekman boundary-layer flow and transport due to an applied tangential stress force at the surface.

Eqs. (7.72), (7.73) imply

$$(\nu \lambda^2)^2 \;=\; -f^2$$

$$\lambda \;=\; +\left(\frac{f}{2\nu}\right)^{\frac{1}{2}}(1 \pm i)\,, \qquad \lambda \;=\; -\left(\frac{f}{2\nu}\right)^{\frac{1}{2}}(1 \pm i)\,. \qquad (7.74)$$

To obtain $u \to u_\infty$, $v \to 0$ as $|z| \to \infty$ we reject the first pair of solutions (7.74) and retain the other two. Thus we obtain

$$u - u_\infty \;=\; e^{-|z|/d}\left(a_0 \cos z/d \,+\, a_1 \sin z/d\right)\,, \qquad (7.75)$$

$$v \;=\; \mathrm{sgn}(z)\,e^{-|z|/d}\left(a_1 \cos z/d \,-\, a_0 \sin z/d\right)\,, \qquad (7.76)$$

where we now define

$$d \;=\; \left(\frac{2\nu}{f}\right)^{\frac{1}{2}}. \qquad (7.77)$$

Since d is the e-folding length over which these solutions vary with distance from the boundary, and hence essentially the boundary layer "thickness", we note that it has the form anticipated in Eq. (7.67).

We now consider two applications. We consider first a solid boundary at $z = 0$, on which $u = v = 0$, with fluid above it ($z > 0$). To match these conditions at $z = 0$ with Eqs. (7.75), (7.76) requires $a_0 = -u_\infty$, $a_1 = 0$. Note that the flow direction spirals as z varies through the boundary layer. Very near $z = 0$ the flow speed is very small but it is at $45°$ to the flow outside the boundary layer.

Next we consider a boundary at $z = 0$ at which a constant stress is applied, providing a constant applied horizontal force $\tau_x e_x + \tau_y e_y$ per unit area. Continuity of stresses at the boundary implies that the corresponding boundary condition on the flow is

$$\rho_0 \nu \frac{\partial u}{\partial z} = \tau_x , \qquad \rho_0 \nu \frac{\partial v}{\partial z} = \tau_y \qquad (7.78)$$

at $z = 0$. Since this is typically found at the upper boundary of a fluid (say the surface of the ocean) with z as height, we consider the fluid to occupy $z < 0$. Thus, taking into account that $-|z|/d \equiv +z/d$ in the exponents in (7.75) and (7.76), it follows that at $z = 0$ $\partial u/\partial z = (a_0 + a_1)/d$ and $\partial v/\partial z = (a_0 - a_1)/d$ and hence, applying (7.78) and using (7.77),

$$a_0 = \frac{\tau_x + \tau_y}{\rho_0 f d} , \qquad a_1 = \frac{\tau_x - \tau_y}{\rho_0 f d} . \qquad (7.79)$$

This implies that at $z = 0$ the current $(u - u_\infty)e_x + ve_y$ arising from the applied stress is at $45°$ to the direction of the applied force. Again, the thickness of the Ekman layer near the boundary is of order $d \sim (\nu/\Omega)^{\frac{1}{2}}$. We can define the driven horizontal transport in the x-direction in the layer:

$$\int_{-\infty}^{0} (u - u_\infty)\rho_0 \, \mathrm{d}z = \frac{1}{f}\tau_y ;$$

and likewise in the y-direction,

$$\int_{-\infty}^{0} v\rho_0 \, \mathrm{d}z = -\frac{1}{f}\tau_x ,$$

after evaluating the integrals using the solutions found above. Therefore, suprisingly perhaps, the so-called Ekman transport is perpendicular to the direction of the applied force (Fig. 7.2).

Chapter 8

Accretion, Winds and Shocks

In this chapter we consider a number of varied but related topics. The first is accretion. Accretion occurs in many astrophysical contexts and on many scales. Examples include: interstellar matter falling onto a star; infall of stars and gas into a black hole in the centre of a galaxy; infall of material into the centre of a rich cluster of galaxies. In the first part of the chapter we therefore consider the case of spherically symmetric accretion whereby matter flows radially onto a central, gravitating mass. The flow is assumed to be steady and viscosity is neglected. The mathematical treatment of the spherically symmetric accretion problem is identical to that of flows away from rather than toward the central mass, specifically spherically symmetric winds from an object such as a star, under the same simplifying assumptions. Therefore our work on the accretion problem enables us without much further effort to say something about stellar winds.

Accretion flows and winds typically give rise to shocks: mathematically one can think of these as necessitated by having to match the flow solutions to boundary conditions at the accreting object and very far away from it, requiring two different mathematical solutions that match at essentially a discontinuity called a shock. In the later part of the chapter we therefore look at some considerations for shocks, and in particular the jump conditions across a shock. We start those considerations with a discussion of steepening acoustic waves an example of shock formation, and end with a discussion of the blast wave from a supernova.

The fluid flow involved in the accretion process may be roughly spherically symmetric; but on the other hand, if the accreting material has significant angular momentum, it cannot just fall radially: rather this leads to the formation of an accretion disk. We defer the important topic of accretion disks until Chapter 9.

8.1   Bernoulli's Theorem

The problem of how material falls radially onto a central object is sometimes called the Bondi problem, after Bondi (1952); see also Shu (1992).

As a preliminary, we prove a standard result in fluid dynamics, Bernoulli's theorem for steady inviscid flow. A standard identity from vector calculus gives

$$\boldsymbol{u} \cdot \nabla \boldsymbol{u} \; = \; \nabla \left(\frac{1}{2} u^2 \right) \; - \; \boldsymbol{u} \times (\nabla \times \boldsymbol{u}) \,, \tag{8.1}$$

where $u = |\boldsymbol{u}|$. Neglecting viscosity and setting time derivatives to zero, and using Eq. (8.1), the equation of motion (2.1) can be written

$$\nabla \left(\frac{1}{2} u^2 \right) \; - \; \boldsymbol{u} \times (\nabla \times \boldsymbol{u}) \; = \; -\frac{1}{\rho} \nabla p \; - \; \nabla \psi \,. \tag{8.2}$$

Suppose that the flow is barotropic, so the pressure is a known function of density, $p = p(\rho)$. This is a common simplification in astrophysical fluid dynamics – by assuming a given relation between pressure and density, we can often avoid needing to give specific consideration to the energy equation. We have already come across barotropic flows in Chapter 7. We define the enthalpy

$$h \; = \; \int \frac{\mathrm{d}p}{\rho} \,, \tag{8.3}$$

so $\nabla h = \rho^{-1} \nabla p$. Then Eq. (8.2) becomes

$$\boldsymbol{u} \times (\nabla \times \boldsymbol{u}) \; = \; \nabla \left(\frac{1}{2} u^2 + h + \psi \right) \,, \tag{8.4}$$

from which it follows by taking the dot product with $\boldsymbol{u}$ that

$$\boldsymbol{u} \cdot \nabla \left(\frac{1}{2} u^2 + h + \psi \right) \; = \; 0 \,. \tag{8.5}$$

This result shows that $\frac{1}{2} u^2 + h + \psi$ is constant along a streamline, i.e. a line everywhere parallel to $\boldsymbol{u}$. This is *Bernoulli's theorem*. (In a steady flow, such as we are considering here, streamlines are also the paths — pathlines — along which material fluid elements travel.)

An everyday application of Bernoulli's theorem is to the flow from a household tap. In this case ρ is essentially uniform, so $h = p/\rho$. Bernoulli's theorem says that

$$\frac{1}{2}u^2 + \frac{p}{\rho} + gz = \text{constant} \tag{8.6}$$

along streamlines, z being measured upwards. Now in particular, along a surface streamline the pressure p is everywhere equal to the atmospheric pressure, which is a constant. Therefore as the flow falls from the tap, Eq. (8.6) implies that u increases. Now suppose that the stream from the tap has horizontal cross-sectional area $A(z)$. The direction of $\boldsymbol{u}$ is essentially vertical, and the flow is incompressible. Thus mass conservation implies that $\rho u A$, the mass flow per unit time through a horizontal plane, is independent of z. Thus, as the water falls, u increases and A decreases.

Note, however, that for sufficiently small A, the surface tension cannot be ignored. Also the flow is not stable: a Kelvin-Helmholtz instability sets up an oscillatory disturbance on the surface.

8.2 The de Laval Nozzle

We consider now how the picture of incompressible-type flow from a tap will be modified in a situation where the compressibility of the fluid is important. We still take the flow to be steady, barotropic and one-dimensional. As an example, consider the flow from a jet engine. The spatial variation of the cross-section A is given (by the walls of the combustion chamber), and we can neglect gravity. Bernoulli and mass conservation imply

$$\frac{1}{2}u^2 + h = \text{constant}$$
$$\rho u A = \text{constant} . \tag{8.7}$$

The spatial variation of A induces variations in the other quantities. The first of Eqs. (8.7) implies that

$$u\,du + \frac{c^2}{\rho}d\rho = 0 , \tag{8.8}$$

where $c^2 = dp/d\rho$ (c is thus the sound speed). This equation, which relates changes in density and in velocity, can be rewritten

$$\frac{d\rho}{\rho} = -\mathcal{M}^2\frac{du}{u} , \tag{8.9}$$

where $\mathcal{M} \equiv u/c$ is the Mach number.

Note that if $\mathcal{M} \ll 1$, fractional changes in density are negligible compared with fractional changes in u. Thus we can generally neglect compressibility if $\mathcal{M} \ll 1$. On the other hand, supersonic flight past obstacles involves substantial compressions and expansions. Also, Eq. (8.9) and the second of Eqs. (8.7) together give

$$(1 - \mathcal{M}^2)\frac{du}{u} = -\frac{dA}{A}. \tag{8.10}$$

We now consider Eq. (8.10) for three cases.

If $\mathcal{M} < 1$ (subsonic flow), an increase in u corresponds to a decrease in A. This was the situation for the running tap.

If $\mathcal{M} > 1$ (supersonic flow), an increase in u requires an increase in the area of the nozzle! The explanation for this, from Eq. (8.9), is that the density decreases faster than the velocity increases; thus mass conservation requires an increase in A.

For $\mathcal{M} = 1$, the sonic transition between subsonic and supersonic flow, for a smooth transition (du finite) Eq. (8.10) implies that dA must be zero at the transition point. This is important for jet design. The nozzle needs to converge (A decreasing) to provide the necessary acceleration from subsonic speeds, but should smoothly stop converging and start to diverge where the flow gets to supersonic speeds. In astrophysical situations, the same acceleration can be achieved by external body forces, such as gravity.

8.3   The Bondi Problem

Now we consider the steady, spherically symmetric accretion of gas onto a gravitating point mass M. We assume a barotropic flow, so $p = p(\rho)$. Also we neglect the self-gravity of the infalling gas, which is a good approximation if its total mass is much less than that of the central point mass.

The velocity is wholly in the inward radial direction. Since the flow is steady, integrating the continuity equation over the region between concentric spherical surfaces and using the divergence theorem gives the mass conservation equation

$$4\pi r^2 \rho u = \text{constant} = -\dot{M} \tag{8.11}$$

where $\dot{M}$ is a positive constant. Bernoulli's theorem (8.6) yields

$$\frac{1}{2}u^2 + h - \frac{GM}{r} = 0 \qquad (8.12)$$

where

$$h = \int_{\rho_\infty}^{\rho} \frac{\mathrm{d}p}{\rho} , \qquad (8.13)$$

ρ_∞ being the density at infinity. Note that Bernoulli's theorem applies to a given radial streamline, and following a streamline out to infinity shows that the constant on the right-hand side of (8.12) is zero: for at infinity $u = 0$, and $h = 0$ there by Eq. (8.13). Since every point in space is on some radial streamline, and the constant is zero on each one of them, Eq. (8.12) holds not just on a single streamline but everywhere in space.

In the particular case of isothermal flow, $p = c_\infty^2 \rho$ where c_∞ is a constant. Evaluating Eq. (8.13) gives

$$h = c_\infty^2 \ln(\rho/\rho_\infty) . \qquad (8.14)$$

In the case of polytropic flow, for which $p = p_\infty(\rho/\rho_\infty)^\gamma$ (p_∞ and γ being constants), Eq. (8.13) yields

$$h = \frac{\gamma}{\gamma - 1} c_\infty^2 \left[\left(\frac{\rho}{\rho_\infty} \right)^{\gamma - 1} - 1 \right] \qquad (8.15)$$

where $c_\infty^2 = p_\infty/\rho_\infty$. In the terminology of Shu (1992), c_∞ is the "thermal speed", while

$$c^2 \equiv \frac{\mathrm{d}p}{\mathrm{d}\rho} = \gamma c_\infty^2 \left(\frac{\rho}{\rho_\infty} \right)^{\gamma - 1} \qquad (8.16)$$

is the square of the "acoustic speed".

A characteristic length is

$$r_B = \frac{GM}{c_\infty^2} . \qquad (8.17)$$

This is called the Bondi radius.

We may define a dimensionless radial variable x, speed $\bar{u}$ and density $\bar{\rho}$ by

$$x = \frac{r}{r_B} , \quad \bar{u} = \frac{|u|}{c_\infty} , \quad \bar{\rho} = \frac{\rho}{\rho_\infty} \qquad (8.18)$$

and a dimensionless accretion rate λ by measuring $\dot{M}$ in units of a mass flux $\rho_\infty c_\infty$ across an area $4\pi r_B^2$:

$$\lambda = \frac{\dot{M}}{4\pi \rho_\infty (GM)^2/c_\infty^3} \, . \tag{8.19}$$

The governing Eqs. (8.11) and (8.12) can then be written in dimensionless form (Shu 1992) as

$$x^2 \bar{\rho}\bar{u} = \lambda \tag{8.20}$$

and

$$\frac{1}{2}\bar{u}^2 + H(\bar{\rho}) - \frac{1}{x} = 0 \tag{8.21}$$

where for $\gamma = 1$ (isothermal flow) $H(\bar{\rho}) = \ln \bar{\rho}$ and for $\gamma \neq 1$ $H(\bar{\rho}) = (\gamma/(\gamma-1))(\bar{\rho}^{\gamma-1} - 1)$.

In the following we shall consider only isothermal flow. Equations (8.20) and (8.21) imply that changes in the different dimensionless variables are related by

$$2\frac{dx}{x} + \frac{d\bar{\rho}}{\bar{\rho}} + \frac{d\bar{u}}{\bar{u}} = 0 \, , \tag{8.22}$$

$$\bar{u}d\bar{u} + \frac{d\bar{\rho}}{\bar{\rho}} + \frac{dx}{x^2} = 0 \, ; \tag{8.23}$$

and eliminating $d\bar{\rho}$ between these gives a relation between dx and $d\bar{u}$:

$$\left(\bar{u} - \frac{1}{\bar{u}}\right)d\bar{u} = \left(\frac{2}{x} - \frac{1}{x^2}\right)dx \, . \tag{8.24}$$

Now we are interested in solutions of Eq. (8.24) that are relevant to the accretion problem. Far from the accreting body the speed of flow is small. With this in mind we obtain solutions such as those illustrated in Fig. 8.1a. Solutions can be labelled according to the value of the accretion rate λ, applying Eq. (8.20) at infinity. There are three qualitatively different kinds of solution, depending on whether λ is less than, greater than, or equal to some critical value λ_c (which we shall define shortly). We reject the solutions for $\lambda > \lambda_c$ as they are double-valued for a given value of x, as can be seen in the figure. This leaves the remaining two classes of solution.

A sonic transition ($\bar{u} = 1$) occurs when $x = 1/2$. (Recall that $\bar{u}$ is the Mach number.) At this point, Eq. (8.21) implies that $\bar{\rho} = \exp(3/2)$ and Eq. (8.20) gives $\lambda = \lambda_c \equiv \frac{1}{4}\exp\left(\frac{3}{2}\right)$. This defines λ_c.

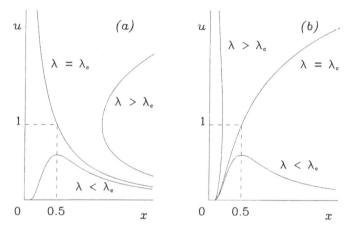

Fig. 8.1   (a) Putative solutions $\bar{u}(x)$ for the isothermal accretion-flow problem, according as the accretion rate λ is smaller than, equal to or larger than its critical value λ_c. All the solutions start at small speed $\bar{u}$ at large distances x. (b) As (a), but for the isothermal wind. All solutions start at small speed $\bar{u}$ at small distances x.

Which solution is the relevant one for a given problem must depend on the boundary conditions. The flows with $\lambda < \lambda_c$ are subsonic everywhere. The flow with $\lambda = \lambda_c$, which has the largest possible mass accretion rate, is subsonic at distances $x > 1/2$ from the star but becomes supersonic for $x < 1/2$. To match onto physical conditions near the accreting body, the flow at some point undergoes a shock transition to a subsonic solution downstream of the shock: see Holzer & Axford (1970).

Now Shu (1992) presents an argument that the solutions with $\lambda < \lambda_c$ are unphysical because they would require unrealistically large densities as $x \to 0$. Whilst this is true of the isothermal solution, real flows will not be isothermal and other physical effects we have not included (heating, for example) will also need to be taken into consideration. Nonetheless the subsonic and critical flows shown in Fig. 8.1a show qualititatively the flows that may be expected.

The maximum accretion rate attainable in the isothermal flow considered above corresponds to $\lambda = \lambda_c$. This implies that

$$\dot{M} \;=\; \lambda_c 4\pi \rho_\infty \frac{(GM)^2}{c_\infty^3} \,. \tag{8.25}$$

This is therefore the maximum rate at which mass can be accreted steadily onto a point mass, assuming spherical symmetry and isothermal flow.

8.4 The Parker Solar-Wind Solution

The same mathematical model we have just derived to describe the Bondi solution for spherically symmetric accretion can be used to describe a steady, spherically symmetric thermally driven stellar wind. The flow is then away from the central mass instead of towards it. This solution was derived by Parker (1958); see also Parker (1963).

Again we restrict our attention to the illustrative isothermal case. The flow is once again described by Eq. (8.24) but now the flow is outwards rather than inwards. The quantity λ will therefore correspond to a mass-loss rate. The flow originates at the star, and so we suppose that the boundary conditions there will require that the velocities be small there. The possible solutions are illustrated in Fig. 8.1b. There are again three qualitatively different solutions: the critical value of λ is the same as before. The flow with $\lambda < \lambda_c$ is subsonic everywhere. The flow corresponding to $\lambda = \lambda_c$ is subsonic for $x < 1/2$, passes through the sonic point and is supersonic for $x > 1/2$. Physical boundary conditions will again generally require that the flow at some point undergoes a shock transition to a subsonic flow downstream (i.e. further away from the star). The solutions are described in more detail in the review by Holzer & Axford (1970), where details of more physically realistic approximations are also considered.

8.5 Nonlinear Acoustic Waves

We have just considered common astrophysical flows describing accretion onto an object or winds flowing away from it, and mentioned that the solutions may develop shock transitions which are essentially discontinuities in the fluid flow. We shall therefore proceed to consider shocks in a little detail. We start with consideration of nonlinear acoustic waves. This serves as a reminder that the linear treatment of much of the rest of this book is not always adequate in astrophysical fluid dynamics. In particular, we show that, taking nonlinear effects into account, acoustic waves steepen rather than propagating without change of form as linear theory predicts. We also look briefly at the powerful method of characteristics.

For simplicity, we consider 1-D waves in a homogeneous medium, with pressure p_0, density ρ_0 and adiabatic sound speed $c_0^2 = \gamma p_0/\rho_0$. Moreover, we consider the motion to be adiabatic, with $p \propto \rho^\gamma$ always. So

$$c^2 = \frac{dp}{d\rho} = \frac{\gamma p}{\rho} = c_0^2 \left(\frac{\rho}{\rho_0}\right)^{\gamma-1} = c_0^2 \left(\frac{p}{p_0}\right)^{1-1/\gamma} . \qquad (8.26)$$

We shall work in a Cartesian geometry, though the generalization to flow depending only on a spherical polar radial coordinate r is straightforward.

Previously, we have considered only the linearized equations

$$\frac{\partial \rho'}{\partial t} + \rho_0 \frac{\partial u}{\partial x} = 0 , \qquad \rho_0 \frac{\partial u}{\partial t} = -\frac{\partial p'}{\partial x} = -c_0^2 \frac{\partial \rho'}{\partial x}$$

from which it follows that

$$\frac{\partial^2 \rho'}{\partial t^2} = c_0^2 \frac{\partial \rho'}{\partial x^2} . \qquad (8.27)$$

This has as its most general solution

$$\begin{aligned}
\rho' &= f(x - c_0 t) + g(x + c_0 t) , \\
p' &= c_0^2 [f(x - c_0 t) + g(x + c_0 t)] , \qquad (8.28) \\
u &= \frac{c_0}{\rho_0} [f(x - c_0 t) - g(x + c_0 t)] .
\end{aligned}$$

Note that individual waveforms maintain their shape forever. If the original disturbance has a finite extent, then the two waves (leftward- and rightward-travelling) become completely separated after a finite time.

We wish now to consider nonlinear effects, as discussed for example in Lighthill (1978). First we consider the following plausibility argument. Consider a rightward-travelling wave (i.e. $g \equiv 0$). Writing $c = c_0 + c'$, it follows from (8.26) that

$$(c_0 + c')^2 = c_0^2 \left(\frac{\rho_0 + \rho'}{\rho_0}\right)^{\gamma-1}$$

and hence

$$\frac{c'}{c_0} \approx \frac{1}{2}(\gamma - 1) \frac{\rho'}{\rho_0} .$$

Thus the sound speed is greater in a more compressed part of the wave. The waveform becomes distorted (Fig. 8.2): the waveform steepens and one may get an unphysical situation where u becomes multivalued.

Consider then the full nonlinear equations:

$$\rho \left(\frac{\partial u}{\partial t} + u \frac{\partial u}{\partial x}\right) = -\frac{\partial p}{\partial x} , \qquad \frac{\partial \rho}{\partial t} + \frac{\partial}{\partial x}(\rho u) = 0 . \qquad (8.29)$$

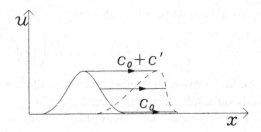

Fig. 8.2   Steepening of an acoustic wave.   Different parts of the waveform travel at
different wave speeds.

We introduce a new variable

$$\mathcal{P} = \int_{p_0}^{p} \frac{\mathrm{d}p}{\rho c} \; ;$$

simple differentiation yields

$$\frac{\partial \mathcal{P}}{\partial x} = \frac{1}{\rho c}\frac{\partial p}{\partial x} \quad \text{and} \quad \frac{\partial \mathcal{P}}{\partial t} = \frac{c}{\rho}\frac{\partial \rho}{\partial t} \, ,$$

where in the last equation we have used Eq. (8.26) to change a pressure
derivative to a density derivative. Now Eqs. (8.29), multiplying the second
by c/ρ, can be written in terms of $\mathcal{P}$ as

$$\frac{\partial u}{\partial t} + u\frac{\partial u}{\partial x} + c\frac{\partial \mathcal{P}}{\partial x} = 0 \, ,$$

$$\frac{\partial \mathcal{P}}{\partial t} + u\frac{\partial \mathcal{P}}{\partial x} + c\frac{\partial u}{\partial x} = 0 \, . \qquad (8.30)$$

Adding and subtracting the equations (8.30) gives

$$\frac{\partial}{\partial t}(u + \mathcal{P}) + (u + c)\frac{\partial}{\partial x}(u + \mathcal{P}) = 0$$

$$\frac{\partial}{\partial t}(u - \mathcal{P}) + (u - c)\frac{\partial}{\partial x}(u - \mathcal{P}) = 0 \qquad (8.31)$$

(e.g. Lighthill 1978). This trick is due to Riemann.

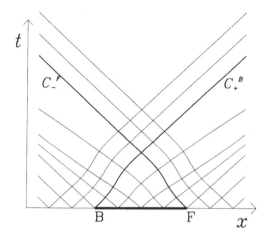

Fig. 8.3   A space-time diagram showing characteristics emanating from a disturbance initially confined in the region between $x = B$ and $x = F$. The C_- characteristic emanating from $x = F$ (C_-^F)and the C_+ characteristic emanating from $x = B$ (C_+^B) are shown as thicker curves.

We introduce a family of curves C_+ in the (x, t) plane defined by $dx = (u+c)dt$, and another family of curves C_- defined by $dx = (u-c)dt$. Then, by Eqs. (8.31), $u + \mathcal{P}$ is constant on each C_+ curve, and $u - \mathcal{P}$ is constant on each C_- curve. These families of curves are called *characteristics*.

As an example of the application of $C_\pm$ curves, we consider the case of a disturbance that is initially confined between $x = B$ and $x = F$ (see Fig. 8.3). We follow closely the treatment by Lighthill. Note the following salient points.

C_- curves to the left of C_-^B (i.e. the C_- characteristic emanating from $x = B$) or to the right of C_-^F originate on the x-axis at $t = 0$ with $u = \mathcal{P} = 0$. Since $u - \mathcal{P}$ is constant on these curves, it follows that $u = \mathcal{P}$ on them.

Likewise, $u + \mathcal{P} = 0$, i.e. $u = -\mathcal{P}$, on C_+ curves to the left of C_+^B or to the right of C_+^F.

Now the whole region to the right of C_+^F is also to the right of C_-^F. Hence $u = \mathcal{P}$ and $u = -\mathcal{P}$ there, i.e. $u = \mathcal{P} = 0$. Thus there is no disturbance in that region. Likewise, there is no disturbance in the region to the left of C_-^B. This is because the disturbance has not had time to propagate into these two regions of the space-time diagram (cf. light-cones in special relativity).

There is also a region to the left of C_+^B and to the right of C_-^F where, by the same argument, $u = \mathcal{P} = 0$. The reason is that the initial disturbance disentangles itself into leftward- and rightward-propagating waves, leaving an undisturbed region in between. We have already noted the analogous behaviour in the linear case.

The rightward-propagating simple wave exists between C_+^B and C_+^F. In the part of this region that is to the right of C_-^F, $u = \mathcal{P}$. But on each C_+ curve, $u + \mathcal{P}$ is constant. Therefore u and $\mathcal{P}$ are individually constant on each C_+ curve. Thus p is constant; and so is c. So $u + c$ is constant: therefore the C_+ characteristics in this region are straight lines, on which u and p' take constant values. Likewise, between C_-^B and C_-^F (to the left of C_+^B), C_- characteristics are straight lines with u and p' constant on them. Thus for a rightward-propagating simple wave,

$$u = \mathcal{P} = \int_{p_0}^{p} \frac{\mathrm{d}p}{\rho c} .$$

One can show that for $p \propto \rho^\gamma$ and $c^2 = c_0^2 (\rho/\rho_0)^{\gamma-1}$,

$$\mathcal{P} = \frac{2(c - c_0)}{\gamma - 1}$$

and so

$$c = c_0 + \frac{1}{2}(\gamma - 1)\mathcal{P} = c_0 + \frac{1}{2}(\gamma - 1)u . \tag{8.32}$$

Thus all quantities are constant on C_+ curves, given by $\mathrm{d}x = (u + c)\mathrm{d}t$. The signal propagates with speed $u + c$ given by

$$u + c = c_0 + \frac{1}{2}(\gamma - 1)u .$$

In particular, therefore, regions with greater speed u propagate more quickly than regions with smaller u, and the wave steepens.

As an example of the development of such a nonlinear wave, consider a piston accelerated from rate. Figure 8.4a shows the case of the piston accelerating *away* from the fluid. The C_+ curves do not cross. By contrast, if the piston accelerates *into* the fluid (Fig. 8.4b) the C_+ curves *do* cross, and u becomes multivalued. Actually this would be unphysical. The description used so far breaks down: physically the fluid develops a shock wave.

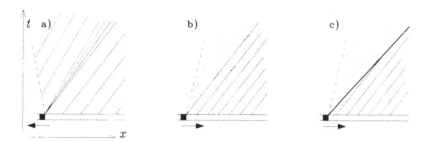

Fig. 8.4   Space-time diagrams showing C_+ characteristics for a 1-D flow caused by the motion of a piston starting impulsively from rest. The space-time line of the piston is shown as the dashed line. (a) The piston moves away from the fluid. (b) The piston moves into the fluid; the characteristics are shown crossing. (c) As (b), but showing the formation of a shock wave (bold line) at the locus of points where the characteristics first cross.

8.6   Shock Waves

The above description predicts an unphysical multivalued velocity u where characteristics cross. In fact, before this happens, large gradients $\partial u/\partial x$ build up. Viscosity, which we have neglected so far, is important in these regions. In reality, the situation is more like that shown in Fig. 8.4c. At the shock front, one passes discontinuously from one C_+ characteristic to another.

To a first approximation, the shock is a discontinuity.

Suppose the shock propagates with steady velocity $u_{\rm sh}$. Ahead of the shock, $\rho = \rho_0$, $u = 0$, $p = p_0$ and $T = T_0$. Behind the shock, $\rho = \rho_1$, $u = u_1$, $p = p_1$, $T = T_1$. Consider a control volume centred on the shock front and moving with it. Mass conservation implies that

$$\rho_1 \left(u_{\rm sh} - u_1 \right) = \rho_0 u_{\rm sh} . \tag{8.33}$$

The change in momentum per unit area and per unit time is given by the pressure difference. Per unit time and unit area, the shock passes through a mass $\rho_0 u_{\rm sh}$ of previously unshocked material. (In the frame of the shock, mass $\rho_0 u_{\rm sh}$ passes through *it*.) In the rest frame, this mass of previously unshocked material acquires momentum $\rho_0 u_{\rm sh} u_1$. This must equal the force per unit area, $p_1 - p_0$. Thus

$$\rho_0 u_{\rm sh} u_1 = p_1 - p_0 . \tag{8.34}$$

These two relations yield u_{sh} and u_1:

$$u_{\text{sh}}^2 = \frac{\rho_1}{\rho_0} \frac{(p_1 - p_0)}{(\rho_1 - \rho_0)} \; ; \quad u_1^2 = \frac{(p_1 - p_0)(\rho_1 - \rho_0)}{\rho_1 \rho_0} . \tag{8.35}$$

Let

$$\epsilon = \frac{p_1 - p_0}{p_0} . \tag{8.36}$$

For a *weak shock*, $\epsilon \ll 1$. In this limiting case, the entropy jump across the shock can be neglected to a first approximation (see below). We can thus assume the same $p(\rho)$ relation both sides of shock so that, for example, $p_1 - p_0 = (\rho_1 - \rho_0)\, \mathrm{d}p/\mathrm{d}\rho$. It is then straightforward to show that

$$u_{\text{sh}}^2 \approx \frac{\rho_1}{\rho_0} \frac{\mathrm{d}p}{\mathrm{d}\rho} \approx c_0^2 ,$$

i.e. $u_{\text{sh}} \approx c_0$, and that $u_1 \approx (p_1 - p_0)/\rho_0 c_0$.

By considering the change in energy of material passing through the shock, one can obtain relations between the pressures and between the densities either side of the shock. For a weak shock in an ideal gas, one can show that the entropy jump $c_V \ln(p_1 \rho_0^\gamma / p_0 \rho_1^\gamma)$ across the shock is $O(\epsilon^3)$, which justifies neglecting the entropy jump in the above argument. One can also show that the thickness of the shock is of order $\nu/\epsilon c_0$, where ν is the kinematic viscosity. Hence we may note that the thickness *decreases* with shock strength. Also, since $\nu \simeq l c_0$, where l is the particle mean free path, the thickness of the shock is of the order of the mean free path and hence the shock is indeed a nearly discontinuous transition. See e.g. Zel'dovich & Raizer (2002).

8.7 Blast Wave from a Supernova

As a final topic in this chapter, we consider the expansion of a blast wave from a supernova explosion. As well as the topic's intrinsic astrophysical interest, the treatment is of interest as an example of the use of a similarity solution: see Shu (1992) for a more detailed discussion.

Consider the supernova as the release of enormous energy E from a point source into an ambient medium of homogeneous density ρ_0, which creates a spherical shock wave, propagating with speed u_{sh}. In the early stages of shock wave development ($\lesssim 10^5$ years), energy is roughly conserved (radiative losses $\ll E$). The ram pressure ρu_{sh}^2 is much greater than the ambient pressure p_0, so we can approximate $p_0 = 0$.

What is the radial position r_{sh} of the shock front at time t?

The only dimensional quantities are E, ρ_0 and t (not u_{sh} — as we shall see below, this is a dependent quantity expressible in terms of the others). Thus we seek to express r_{sh} as a function of the other variables:

$$r_{sh} = E^\alpha \rho^\beta t^\gamma , \tag{8.37}$$

where α, β and γ are constants to be determined. By expressing each of the quantities in (8.37) in terms of dimensions M (mass), L (length) and T (time), Eq. (8.37) implies

$$L = (ML^2T^{-2})^\alpha (ML^{-3})^\beta T^\gamma ; \tag{8.38}$$

whence $\alpha = 1/5$, $\beta = -1/5$ and $\gamma = 2/5$. Thus

$$r_{sh} = A \left(Et^2/\rho_0\right)^{1/5} , \tag{8.39}$$

where A is a dimensionless constant. (From a numerical calculation, with $\gamma = 5/3$, it is found that $A = 1.17$.)

It follows from Eq. (8.39) that

$$u_{sh} = \frac{dr_{sh}}{dt} = \frac{2}{5}A \left(E/\rho_0 t^3\right)^{1/5} . \tag{8.40}$$

From Eqs. (8.39) and (8.40) we can estimate the size and speed of a supernova shock wave as a function of time after the initial explosion. We take as representative values $\rho_0 = 2 \times 10^{-21}$ kg m^{-3} (10^6 atoms per cubic metre) and $E = 10^{44}$ J (corresponding to one solar mass ejected at 10^4 km s^{-1}). After one year, the shock wave has propagated a distance of 0.3 parsecs (pc), and is travelling at 130 000 km s^{-1}. After 10 years, its radius is 0.8 pc and its speed is 32 000 km s^{-1}. After 100 years, its radius is 2 pc and its speed is 8 000 km s^{-1}. After 1 000 years, its radius is 5 pc and its speed is 2 000 km s^{-1}; while after 10 000 years, its radius is 13 pc and its speed is 5 00 km s^{-1}. One can also show with these data that the temperature would be $T \simeq 3 \times 10^6$ K at $t = 10^4$ years, which implies that the gas should be emitting strongly in the X-ray band. Indeed, one does observe supernova remnants in X-ray with sizes of order 10 pc.

Chapter 9

Viscous Accretion Disks

Accretion disks, systems in which a central gravitating mass accretes material from a surrounding disk, appear to be common in astrophysics. Examples of where accretion disks are believed to occur are: when mass transfer takes place onto a black hole or neutron star from a binary companion (an X-ray binary); likewise onto a white dwarf (a nova or dwarf nova/cataclysmic variable); and, it is believed, in quasars (with a massive central black hole). For a very readable account of accretion disks in astrophysics, see the review article by Pringle (1981).

A disk can be characterized by its typical thickness, H, and its radius R. There are various levels of complexity in modelling disks, depending upon whether one considers thick or thin disks (depending on the ratio H/R), and also whether or not the self-gravitation of the accreting matter is taken into account. We shall consider the simplest scenario, namely a thin disk ($H \ll R$) and neglecting the self-gravity of the infalling gas.

9.1 Role of Angular Momentum and Energetics of Accretion

In Chapter 8 the radial infall of accreting matter was considered. If the matter has non-negligible angular momentum, however, it will not fall radially inwards. Suppose the infalling matter has angular momentum h per unit mass, and the central object has mass M and radius R. If the matter falls without loss of angular momentum, then if it were to reach the surface of the central body it would have angular velocity $\Omega = h/R^2$ (since $h = \Omega R^2$). Its centrifugal acceleration would then be $\Omega^2 R = h^2/R^3$, while the gravitational acceleraton would be GM/R^2. Clearly the material cannot fall onto the central object if the former is greater than the latter. Thus spherically symmetric accretion is only a good approximation if $h^2/R^3 \ll GM/R^2$. If $h \gtrsim (GMR)^{1/2}$, one may instead expect the matter to form an accretion disk.

A particle in circular orbit around a central mass will stay there. If energy, and hence angular momentum, is extracted, the particle can spiral in toward the centre. The gravitational potential energy that a mass m loses in falling from a great distance down to the surface of the central body is, using the same notation as before, GMm/R. For a massive compact central object, this can be a sizeable fraction of the infalling matter's rest mass energy, mc^2. For example, for a neutron star of mass $M = 1M_\odot$ (2×10^{30} kg) and $R = 10$ km, the gravitational potential energy lost is about 15% of the rest mass energy. For a central black hole, using for R the Schwarzschild radius $R_s = 2GM/c^2$, the figure is 50%.

The basic idea of a viscous accretion disk is that energy and angular momentum can be removed from gaseous material in the accretion disk by viscosity or some similar dissipative process. The gas loses gravitational energy as it spirals in, and this is converted to heat by dissipation. Provided the gas radiates this heat away efficiently, gravitational energy is thus converted to radiation. The large proportions of the rest mass energy that can thus be released as radiation make accretion disks attractive explanations of highly energetic astrophysical sources.

Actually, the figure of 50% for a black hole is an overestimate of the energy that can be extracted. The material will in fact only continue to spiral in slowly until it reaches the smallest stable circular orbit around the black hole, the radius of which is somewhat greater than R_s. Thereafter the matter will fall in on a much shorter timescale, probably too quickly for the matter to radiate away the gravitational energy it is losing: see Eardley & Press (1975) for details. The proportion of rest mass energy that can be radiated prior to this can still be large, up to 40%. This incidentally indicates why accretion onto a black hole via a viscous disk is a more attractive candidate for a highly energetic source than spherically symmetric accretion onto a similar orbit: there is much more time for the heat to be radiated as the matter slowly spirals in than if the gas just fell radially inwards.

As stated above, here we shall only consider thin disks. in which the thickness is much smaller than the disk radius.

9.2 Thin Accretion Disks

We assume that the gas can radiate efficiently, and that the viscous timescale (on which viscosity can redistribute angular momentum) is longer than other timescales. The infalling gas initially loses as much energy as possible in collisions, while maintaining its angular momentum, and settles into circular orbits, since these are the orbits of least energy. In modelling the disk thus formed, we shall use cylindrical polar coordinates (ϖ, ϕ, z). We assume that the disk is thin and axisymmetric, that $\partial/\partial\phi \equiv 0$ and that $\boldsymbol{u} \cdot \boldsymbol{e}_z = 0$ (no velocity out of the plane of the disk).

Writing the fluid velocity as

$$\boldsymbol{u} = (u, \, \varpi\Omega, \, 0) \,, \tag{9.1}$$

the radial component of the momentum equation (2.1) is

$$\frac{\partial u}{\partial t} + u\frac{\partial u}{\partial \varpi} - \varpi\Omega^2 = -\frac{1}{\rho}\frac{\partial p}{\partial \varpi} - \frac{\partial \psi}{\partial \varpi} \,. \tag{9.2}$$

We assume that the radial pressure gradient is negligible compared with centrifugal forces. Then, with

$$\psi = -\frac{GM}{(\varpi^2 + z^2)^{1/2}} \,, \tag{9.3}$$

to a first approximation $u = 0$ and Eq. (9.2) in the plane of the disk ($z = 0$) yields

$$\varpi\Omega^2 = \frac{GM}{\varpi^2} \,, \tag{9.4}$$

a balance between centrifugal and gravitational forces. This gives

$$\Omega \propto \varpi^{-3/2} \,, \tag{9.5}$$

which is called Keplerian rotation: note that Ω is a function only of ϖ.

In the absence of motion in the z-direction, the z-component of the momentum equation (2.1) gives just

$$\frac{1}{\rho}\frac{\partial p}{\partial z} = -\frac{\partial \psi}{\partial z} \equiv \frac{\partial}{\partial z}\left[\frac{GM}{(\varpi^2 + z^2)^{1/2}}\right] \,, \tag{9.6}$$

i.e. the disk is in hydrostatic equilibrium in the z-direction. Since the disk is thin, z is small so we can use a binomial expansion to approximate (9.6) as

$$\frac{1}{\rho}\frac{\partial p}{\partial z} \approx -\frac{GMz}{\varpi^3} \, . \tag{9.7}$$

As an order-of-magnitude estimate, we can approximate $|z|$ by H and $|\partial p/\partial z|$ by p/H. Hence, in terms of the isothermal sound speed $a \equiv (p/\rho)^{1/2}$, and using Eq. (9.4), Eq. (9.7) yields

$$H \simeq \frac{a}{\varpi\Omega}\varpi \, . \tag{9.8}$$

Hence the thin-disk approximation, that H is small compared to ϖ, is valid provided the isothermal sound speed is much smaller than the rotational velocity of the disk. This is consistent with our assumption that the disk can radiate heat efficiently, so the disk is 'cold' — cf. Eq. (2.12).

Henceforth it will be convenient to work in the approximation of an infinitely thin disk lying in the $z = 0$ plane, so we shall write the density in terms of a surface density $\Sigma(\varpi, t)$:

$$\rho = \Sigma(\varpi, t)\delta(z) \, . \tag{9.9}$$

The *shear* is defined to be

$$A = \varpi\frac{d\Omega}{d\varpi} \, . \tag{9.10}$$

Note that for a rigidly rotating disk, $d\Omega/d\varpi = 0$, so there is no shear – so no "friction" between neighbouring rings in the disk. But in a differentially rotating disk, $A \neq 0$; in particular, for Keplerian rotation, $A \propto \varpi^{-3/2}$.

Gas in different orbits moves at different speeds. If viscosity is present, there will be friction between neighbouring rings of material, causing a redistribution of energy and angular momentum. The friction releases energy as heat, which gets radiated away. The energy the gas loses must come from its gravitational potential energy, and thus it falls further into the potential well of the central object. Thus viscosity converts potential energy into radiation. In the next section we shall make this more quantitative.

According to the above argument, material losing energy falls to an orbit of smaller radius. But note that in the smaller orbit, the orbital speed and hence the kinetic energy of the gas actually increases. It is straightforward to show, using Eq. (9.4), that the kinetic energy per unit mass of the orbiting material is $GM/2\varpi$, and hence that the total of kinetic plus gravitational potential energy does indeed decrease as the gas moves to smaller radii. A corollary of this result is that only half of the potential energy lost is available to be radiated away, the other half going into orbital kinetic energy. The kinetic energy is finally dissipated as heat in a thin boundary layer as the material finally accretes onto the central object (e.g. neutron star). Hence one expects half of the luminosity to come from this thin boundary layer — see Pringle (1981) for more details.

9.3   Diffusion Equation for Surface Density

We define j to be the angular momentum per unit mass (the specific angular momentum):

$$j = \varpi e_\phi \cdot u = \varpi^2 \Omega . \tag{9.11}$$

One can obtain an equation expressing conservation of angular momentum by taking ϖ times the ϕ-component of the momentum equation (1.7):

$$\frac{\partial j}{\partial t} + u\frac{\partial j}{\partial \varpi} = \varpi e_\phi \cdot f \equiv \varpi f_\phi , \tag{9.12}$$

where f is the viscous force, per unit mass. More generally,

$$\frac{Dj}{Dt} = \varpi f_\phi , \tag{9.13}$$

which states that the rate of change of angular momentum is equal to the torque acting.

The viscous force in the i-direction on a plane whose normal is in the j-direction is $\sigma_{ij} = e_i \cdot \sigma \cdot e_j$, where σ is the stress tensor. So for example, if in a Cartesian coordinate system the velocity is u_2 in the x_2-direction and depends only on coordinate x_1, then by Eq. (1.8) the force in the x_2-direction on a plane with normal in the x_1-direction is $\sigma_{12} = \mu du_2/dx_1$. We have a similar situation, namely a velocity $u = \varpi\Omega(\varpi)e_\phi \equiv u_\phi e_\phi$ in the ϕ-direction, depending only on coordinate ϖ. By Eq. (1.8), the viscous force in the ϕ-direction on a surface of constant ϖ is therefore

$$e_\phi \cdot \sigma \cdot e_\varpi = \mu\left[e_\phi \cdot (e_\varpi \cdot \nabla(u_\phi e_\phi)) + e_\varpi \cdot (e_\phi \cdot \nabla(u_\phi e_\phi))\right]$$

$$= \mu\left[e_\phi \cdot \left(\frac{du_\phi}{d\varpi}e_\phi\right) + e_\varpi \cdot \left(-\frac{1}{\varpi}e_\varpi\right)\right] \tag{9.14}$$

$$= \mu\left[\frac{d}{d\varpi}(\varpi\Omega) - \Omega\right] = \mu\varpi\frac{d\Omega}{d\varpi} = \mu A .$$

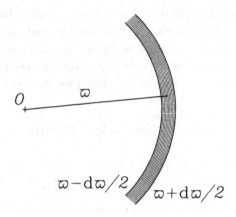

Fig. 9.1   Annulus of material between radii $\varpi - d\varpi/2$ and $\varpi + d\varpi/2$.

Consider then the viscous force on a narrow ring of thickness $d\varpi$, between radii $\varpi - d\varpi/2$ and $\varpi + d\varpi/2$ (Fig. 9.1). For Keplerian rotation, $d\Omega/d\varpi < 0$, so the viscous force due to material at radius less than $\varpi - d\varpi/2$ tends to speed up the ring, and material at radius greater than $\varpi + d\varpi/2$ tends to slow down the ring.

The viscous force per unit area acting on the ring due to material *inside* of the ring, in the sense of speeding it up, is

$$(\mu|A|)_{\varpi - d\varpi/2} \, .$$

Likewise the force per unit area acting on the ring due to material *outside* of the ring, in the sense of slowing it down, is

$$(\mu|A|)_{\varpi + d\varpi/2} \, .$$

Integrating over z, and recalling that $\mu = \rho\nu$, the force due to interior material is

$$F \;=\; 2\pi \left[\varpi\nu|A| \int_{-\infty}^{\infty} \rho \, dz \right]_{\varpi - d\varpi/2} \;=\; 2\pi \left(\varpi\nu\Sigma|A| \right)_{\varpi - d\varpi/2} \, .$$

One can obtain a similar expression for the force due to the exterior material, but evaluated at $\varpi + d\varpi/2$. Torque is just ϖF. Thus the net torque on the ring, in the sense of speeding it up, is

$$-2\pi \left(\varpi^2 \nu \Sigma \varpi \frac{\mathrm{d}\Omega}{\mathrm{d}\varpi} \right)_{\varpi - \mathrm{d}\varpi/2} + 2\pi \left(\varpi^2 \nu \Sigma \varpi \frac{\mathrm{d}\Omega}{\mathrm{d}\varpi} \right)_{\varpi + \mathrm{d}\varpi/2} \qquad (9.15)$$
$$= 2\pi \frac{\mathrm{d}}{\mathrm{d}\varpi} \left(\varpi^3 \nu \Sigma \frac{\mathrm{d}\Omega}{\mathrm{d}\varpi} \right) \mathrm{d}\varpi .$$

The mass of the ring is $2\pi\varpi\Sigma\,\mathrm{d}\varpi$. Putting these two together, the torque ϖf_ϕ per unit mass acting on the ring is

$$\varpi f_\phi = \frac{1}{\varpi\Sigma} \frac{\mathrm{d}}{\mathrm{d}\varpi} \left(\varpi^3 \nu \Sigma \frac{\mathrm{d}\Omega}{\mathrm{d}\varpi} \right) . \qquad (9.16)$$

Thus the angular momentum equation (9.13) becomes

$$\frac{Dj}{Dt} = \frac{1}{\varpi\Sigma} \frac{\mathrm{d}}{\mathrm{d}\varpi} \left(\varpi^3 \nu \Sigma \frac{\mathrm{d}\Omega}{\mathrm{d}\varpi} \right) . \qquad (9.17)$$

Integrating the mass conservation equation (2.2) with respect to z yields a second equation,

$$\frac{\partial\Sigma}{\partial t} + \frac{1}{\varpi} \frac{\partial}{\partial\varpi}(\varpi u\Sigma) = 0 . \qquad (9.18)$$

Since Ω is a function only of ϖ, so is j; hence

$$\frac{Dj}{Dt} = u\frac{\mathrm{d}j}{\mathrm{d}\varpi} . \qquad (9.19)$$

Thus Eqs. (9.17) and (9.18) can be combined to yield

$$\frac{\partial\Sigma}{\partial t} + \frac{1}{\varpi} \frac{\partial}{\partial\varpi} \left[\left(\frac{\mathrm{d}j}{\mathrm{d}\varpi} \right)^{-1} \frac{\mathrm{d}}{\mathrm{d}\varpi} \left(\varpi^3 \nu \Sigma \frac{\mathrm{d}\Omega}{\mathrm{d}\varpi} \right) \right] = 0 \qquad (9.20)$$

or, eliminating j,

$$\frac{\partial\Sigma}{\partial t} + \frac{1}{\varpi} \frac{\partial}{\partial\varpi} \left(\left(\frac{\mathrm{d}\varpi^2\Omega}{\mathrm{d}\varpi} \right)^{-1} \frac{\mathrm{d}}{\mathrm{d}\varpi} \left[\varpi^3 \nu \Sigma \frac{\mathrm{d}\Omega}{\mathrm{d}\varpi} \right] \right) = 0 . \qquad (9.21)$$

In particular, for Keplerian rotation (9.5) due to a central point mass,

$$\frac{\partial\Sigma}{\partial t} = \frac{3}{\varpi} \frac{\partial}{\partial\varpi} \left(\varpi^{1/2} \frac{\partial}{\partial\varpi} \left[\nu \Sigma \varpi^{1/2} \right] \right) . \qquad (9.22)$$

In general, ν may be expected to depend on ϖ, t and Σ: hence Eq. (9.22) constitutes a nonlinear diffusion equation for Σ. On the other hand, if ν is just a prescribed function of radius, the diffusion equation is linear; and if ν

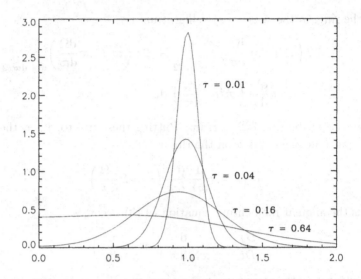

Fig. 9.2   Evolution of the surface density of an annulus of material of mass m according to Eq. (9.22), as a function of dimensionless radius $x = r/r_0$, where r_0 is the initial radius of the initially infinitesimally narrow annulus. The surface density is represented in units of $m/\pi r_0^2$. The various times are indicated in units of a dimensionless time $\tau \equiv 12\nu t/r_0^2$. (After Pringle 1981.)

varies as a power of ϖ, then the equation can be solved analytically. Pringle (1981) discusses the case of constant viscosity ν. Specifically, he illustrates that if the matter is initially concentrated in a ring at some radius, then this diffuses so that at late times most of the mass initially in the ring approaches the central object, while a small fraction of the mass carries most of the angular momentum to large distances (see Fig. 9.2).

9.4   Steady Disks

The problem of modelling the structure of an accretion disk can be simplified still further by seeking time-independent solutions. In that case, the mass conservation equation (9.18) gives

$$2\pi\varpi\Sigma(-u) = \dot{M}_d \qquad (9.23)$$

where $\dot{M}_d$ is some constant. The factor 2π is included here so that $\dot{M}_d$ has the physical interpretation that it is the inward flux of mass per unit time across a circle of radius ϖ. Note that the radial velocity u is negative and $\dot{M}_d$ is positive. Equation (9.16) can be written in terms of $\dot{M}_d$ as

$$-\dot{M}_d \frac{\mathrm{d}j}{\mathrm{d}\varpi} = \frac{\mathrm{d}\mathcal{T}}{\mathrm{d}\varpi}, \qquad (9.24)$$

using Eq. (9.19), where

$$\mathcal{T} = 2\pi\varpi^3 \nu\Sigma \frac{\mathrm{d}\Omega}{\mathrm{d}\varpi}. \qquad (9.25)$$

With the time derivative set to zero, Eq. (9.20) can be integrated once with respect to ϖ to give

$$\frac{\mathrm{d}}{\mathrm{d}\varpi}\left(\nu\Sigma\varpi^3 \frac{\mathrm{d}\Omega}{\mathrm{d}\varpi}\right) = B\frac{\mathrm{d}j}{\mathrm{d}\varpi} = -\frac{\dot{M}_d}{2\pi}\frac{\mathrm{d}j}{\mathrm{d}\varpi}, \qquad (9.26)$$

where the integration constant B has been evaluated using equation (9.24); and integrating a second time yields

$$2\pi\nu\Sigma\varpi^3 \frac{\mathrm{d}\Omega}{\mathrm{d}\varpi} = -\dot{M}_d\varpi^2\Omega + C, \qquad (9.27)$$

where C is a second integration constant. C is usually determined from conditions at the inner boundary of the accretion disk. Suppose, for example, that the azimuthal velocity u_ϕ in the disk is almost everywhere Keplerian, so $\mathrm{d}\Omega/\mathrm{d}\varpi < 0$. This cannot hold all the way to the surface of the central star, however, and so at some point the rotation must slow down (with decreasing ϖ) to match on to the stellar surface rotation, i.e. $\mathrm{d}\Omega/\mathrm{d}\varpi > 0$ close to the star. At some point $\varpi = \varpi_i$ (say), we have $\mathrm{d}\Omega/\mathrm{d}\varpi = 0$. We shall think of $\varpi = \varpi_i$ as the inner boundary of the accretion disk. If the transition from Keplerian to non-Keplerian rotation takes place over a short distance, we can evaluate Ω on the right-hand side of (9.27) at ϖ_i using the Keplerian expression. Thus we get $C \approx \dot{M}_d(GM\varpi_i)^{1/2}$ and Eq. (9.27) may be rewritten as

$$\nu\Sigma = \frac{\dot{M}_d}{3\pi}\left[1 - \left(\frac{\varpi_i}{\varpi}\right)^{1/2}\right]. \qquad (9.28)$$

In Pringle (1981) the above scenario is justified by arguing that there will be a thin boundary layer near the surface of the star where the transition takes place, so $\varpi_i = R$; but see Shu (1992).

To evaluate the viscous dissipation of heat in the disk, we revisit the thermal energy equation (1.19). The second term on the left-hand side of that equation, written in terms of velocity using mass conservation equation (1.4), is $(p/\rho)\nabla \cdot \boldsymbol{u}$. Equation (1.19) assumed an inviscid fluid, however. In the presence of viscosity, this term in the thermal energy equation becomes

$$\rho^{-1}\sigma_{ij}\frac{\partial u_i}{\partial x_j} \equiv \rho^{-1}\sigma_{ij}e_{ij} \tag{9.29}$$

(noting that σ_{ij} is symmetric), where e_{ij} is the rate of strain tensor defined by Eq. (1.52). In the present case, with $\nabla \cdot \boldsymbol{u} = 0$, and using Eq. (1.8), the viscous part of Eq. (9.29) gives the viscous dissipation as

$$2\nu e_{ij}e_{ij} \tag{9.30}$$

per unit mass. Now for velocity field $\boldsymbol{u} = \varpi\Omega(\varpi)\boldsymbol{e}_\phi$, the velocity gradient tensor is $\nabla\boldsymbol{u} = (\varpi\Omega)'\boldsymbol{e}_\varpi\boldsymbol{e}_\phi - \Omega\boldsymbol{e}_\phi\boldsymbol{e}_\varpi$ and so the only non-zero components ($e_{\varpi\phi}$ and $e_{\phi\varpi}$) of the rate of strain tensor are equal to $(1/2)\varpi\Omega'$, where the prime denotes the derivative with respect to ϖ; from whence, multiplying (9.30) by Σ gives the heat dissipation per unit area per unit time:

$$D(\varpi) = \nu\Sigma\left(\varpi\frac{d\Omega}{d\varpi}\right)^2. \tag{9.31}$$

Substituting from Eq. (9.28) gives

$$D(\varpi) = \frac{3GM\dot{M}_d}{4\pi\varpi^3}\left[1 - \left(\frac{\varpi_i}{\varpi}\right)^{1/2}\right]. \tag{9.32}$$

A convenient feature of this result is that it is independent of the viscosity, which is a major uncertainty in the theory of viscous accretion disks. This is hardly surprising, however, since the energetics are determined by the potential energy loss and the assumption of Keplerian rotation, as outlined above. Multiplying Eq. (9.32) by $2\pi\varpi$ and integrating with respect to radius gives the luminosity from the disk:

$$L = \int_{\varpi_i}^{\infty} 2\pi\varpi\, D(\varpi)\,d\varpi = \frac{1}{2}\frac{GM\dot{M}_d}{\varpi_i}. \tag{9.33}$$

This is only half of the total potential energy loss: the other half goes into rotational kinetic energy of the disk.

Note that Eq. (9.28), together with Eq. (9.23), implies that $u = -3\nu/2\varpi$ and hence $u = O(\nu/\varpi)$, a result that might have been anticipated on dimensional grounds.

There has been quite a considerable discussion in the literature about the question of the stability of steady disk models. However, as Pringle (1981) points out, an instability of a disk model does not necessarily correspond to a physically signficant instability. It should most likely just be interpreted as indicating that there is an inconsistency in the assumptions going in to building the steady model. See Pringle (1981) for some further discussion.

9.5   The Need for Anomalous Viscosity

We have seen that steady accretion disks provide an attractive explanation for various highly energetic sources, but we have not considered the nature of the viscosity. From Eq. (9.21), one can estimate the timescale t_ν for a disk to reach a steady state:

$$\frac{\Sigma}{t_\nu} \simeq \frac{1}{\varpi^2}(\varpi\Omega)^{-1}\frac{\varpi^3\Omega\nu\Sigma}{\varpi^2} \simeq \frac{\nu\Sigma}{\varpi^2}\,, \tag{9.34}$$

so $t_\nu \simeq \varpi^2/\nu$. If ν is given by the molecular viscosity (cf. Section 2.8.2), then one obtains a timescale t_ν longer than the age of the universe.

Alternatively, from Eq. (9.24) one can obtain

$$\dot{M}_d \simeq \nu\Sigma\,. \tag{9.35}$$

Molecular viscosity gives a negligible value for $\dot{M}_d$, yet objects need to be able to accrete the matter that is being dumped on them.

Hence it is believed that there must be a source of *anomalous viscosity*, i.e. some physical mechanism that has an effect equivalent to that of viscosity. There have been various proposals in the literature as to what this anomalous viscosity might be: see Pringle (1981) and Shu (1992). Small-scale turbulence could have an effect on the large-scale motion similar to the effect of viscosity; or maybe magnetic fields could provide the necessary viscosity.

As a final remark on accretion disks, note that we have only considered the case of a thin disk. If the radiative transfer is inefficient, then the disk will not be thin (and *vice versa*). In that case, $H \simeq \varpi$, and the problem of modelling the disk structure is very much analogous to modelling the interior of a star.

Chapter 10

Jeans Instability and Star Formation

Stars are a fundamental element of astrophysics, and understanding how stars form is a fundamentally important question. One interest in star formation is relatively how many stars of different masses are formed (the so-called initial mass function). Since the "light-to-mass ratio" of stars is a function of mass (more massive stars are *much* more luminous than lower mass stars) converting observed luminosities of stellar systems to masses requires among other things an estimate of the mix of stellar masses present; and in particular, a substantial fraction of the mass of stars could be in the form of intrinsically very faint low-mass stars, e.g. brown dwarfs.

Stars form from interstellar clouds of gas. Questions in star formation include whether or not low- and high-mass stars form by the same mechanism, and at the same time and place; and whether star formation is a spontaneous process or requires an external trigger such as a nearby supernova explosion or the passage through the interstellar medium of the galaxy of a spiral density wave. Rotation and magnetic fields, and the formation of disks around young stars, are likely important factors. The subject of star formation is very active and there is a large literature. Further reading may be found in the series of volumes on protostars and planets (e.g. Mannings *et al.* 2000); see also e.g. Shu *et al.* (1987) and Lada & Kylafis (1991). Much of the theoretical work on star formation involves computational simulation. Here however we shall focus on just a few classical analytical ideas pertaining to star formation, and specifically the Jeans gravitational instability. We shall return to a brief review of some of the more recent ideas and results towards the end of the chapter.

10.1　Links to Observations

A body of observational evidence links star formation to interstellar gas clouds: associations of young, bright massive stars are found in nebulae; nebulosity is seen in young, open clusters; and infrared observations reveal young stellar objects (YSOs) obscured by gas.

Note that by the virial theorem (1.44), assuming that the velocity and time derivatives are approximately zero,

$$3 \int p \, dV + \Psi = 0 \, . \tag{10.1}$$

For a monatomic perfect gas, the internal energy per unit mass $U = (3/2)p/\rho$; more generally, for constant γ, $U = (\gamma - 1)^{-1}p/\rho$. Hence

$$3(\gamma - 1)U_{\text{tot}} + \Psi = 0 \, , \tag{10.2}$$

where $U_{\text{tot}} = \int U \rho dV$. Take for example $\gamma = 5/3$. Then $U_{\text{tot}} = -\Psi/2$. Therefore the total (internal plus gravitational) energy $U_{\text{tot}} + \Psi = \Psi/2$ (< 0). As a protostar collapses, Ψ becomes more negative, as does the total energy; but the internal energy (and hence, in general, the temperature) increases. Half the gravitational energy lost goes into internal energy, half gets radiated away (cf. Section 9.2): thus the objects are seen as YSOs in the infrared.

Indeed, because stars are born shrouded in gas and dust it is essential to search for the youngest stellar objects using infrared astronomy. A landmark infrared observatory was the Infrared Astronomical Satellite (IRAS): see the review by Beichman (1987) for a survey of the results from IRAS within our own galaxy, including star formation.

We turn now to one of the fundamental theoretical ideas, that a sufficiently massive gas cloud will not be gravitationally stable but will be liable to the Jeans instability.

10.2　Jeans Instability

Our starting point is the set of linearized equations for small perturbations: Eqs. (2.32)–(2.34). We consider the simplest possible system, which is a homogeneous cloud, infinite in all directions, so p_0 and ρ_0 are independent of position, as too is ψ_0 by virtue of the hydrostatic equation. Thus Eq. (2.32) becomes

$$\rho_0 \frac{\partial \boldsymbol{u}}{\partial t} = -\nabla p' - \rho_0 \nabla \psi' . \tag{10.3}$$

Taking the divergence of this equation, and using Eq. (2.34) to eliminate $\nabla^2 \psi'$, gives

$$\rho_0 \frac{\partial (\nabla \cdot \boldsymbol{u})}{\partial t} = -\nabla^2 p' - 4\pi G \rho_0 \rho' . \tag{10.4}$$

In a uniform medium, $q' = \delta q$ for any quantity q. Suppose that the gas is ideal and isothermal, so $\delta p / p = \delta \rho / \rho$. Hence

$$p' = a^2 \rho' , \tag{10.5}$$

where $a = \sqrt{p_0/\rho_0}$ is the isothermal sound speed. Also in a uniform medium Eq. (2.33) becomes

$$\frac{\partial \rho'}{\partial t} = -\rho_0 \nabla \cdot \boldsymbol{u} . \tag{10.6}$$

Hence, using Eqs. (10.5) and (10.6) to eliminate p' and $\nabla \cdot \boldsymbol{u}$ from (10.4), we obtain

$$\frac{\partial^2 \rho'}{\partial t^2} = a^2 \nabla^2 \rho' + 4\pi G \rho_0 \rho' . \tag{10.7}$$

This is a linear PDE for ρ', with coefficients that are independent of position and time. Hence we may seek solutions $\rho' \propto \exp(i\boldsymbol{k} \cdot \boldsymbol{x} - i\omega t)$. In this case $\partial \rho'/\partial t \equiv -i\omega \rho' \times$ and $\nabla \rho' \equiv i\boldsymbol{k}\rho' \times$. Hence, Eq. (10.7) implies

$$-\omega^2 \rho' = -k^2 a^2 \rho' + 4\pi G \rho_0 \rho' , \tag{10.8}$$

where $k = |\boldsymbol{k}|$; and so for a nontrivial solution ($\rho' \neq 0$) we obtain the dispersion relation

$$\omega^2 = k^2 a^2 - 4\pi G \rho_0 . \tag{10.9}$$

If $\boldsymbol{k}$ and ω are real, this represents an oscillation. However, if the right-hand side of (10.9) is negative, as it will be for sufficiently small k, then ω^2 will be negative and so the cloud will be unstable because there will be a solution ρ' where $|\exp(i\omega t)|$ grows exponentially with time. Thus the cloud is unstable to fluctuations of wavenumber $\boldsymbol{k}$ if

$$k^2 a^2 < 4\pi G \rho_0 ,$$

i.e. if

$$k^{-1} > \left(\frac{a^2}{4\pi G\rho_0}\right)^{1/2} \equiv (2\pi)^{-1}\lambda_J \,.$$

Now a real cloud is of finite size, so one cannot have arbitrarily large wavelengths λ, i.e. arbitrarily small wavenumbers. If the cloud is roughly spherical with radius R, one must have $\lambda \lesssim 2R$. Such a cloud can accommodate perturbations of sufficiently large wavelength and hence be unstable to density perturbations if

$$R \gtrsim R_J \equiv \frac{1}{2}\lambda_J = \frac{1}{2}\left(\frac{\pi a^2}{G\rho_0}\right)^{1/2} \,. \tag{10.10}$$

Put another way, the cloud is unstable and collapses if its mass exceeds the critical Jeans mass

$$M_J = \frac{4}{3}\pi R_J^3 \rho_0 = \frac{\pi}{6}\rho_0\lambda_J^3 \,. \tag{10.11}$$

Note the crucial role of the perturbation to the gravitational potential in this instability. If this had been neglected in Eq. (10.3), Eq. (10.9) would have reduced to $\omega^2 = k^2 a^2$, the dispersion relation for isothermal sound waves, Eq. (2.51), which would give stable oscillatory solutions always.

10.3 Jeans Instability with Rotation

Obviously, although the model of a homogenous cloud has led us to a useful criterion (the Jeans mass) for the collapse of a cloud under its own self-gravity, the model is too simplistic to explain in detail the formation of real stars. A different (albeit very idealized) model is to consider the stability of an infinite thin sheet or disk. The linearized perturbation equations for a thin disk are presented in, e.g. Shu (1992). If one then considers plane-wave perturbations proportional to $\exp(ikx - i\omega t)$, one finds (Larsen 1985) that the dispersion relation for such perturbations is

$$-\omega^2 = 2\pi G\Sigma k - k^2 c^2 \,, \tag{10.12}$$

where Σ is the surface density (mass per unit area) of the disk. Unlike the homogeneous cloud, whose perturbations grew most quickly for the smallest k, the fastest-growing perturbations in the thin-sheet model correspond to a finite $k = \pi G\Sigma/c^2$.

A criterion for the collapse of a long, elongated (cigar-shaped) cloud can be obtained through a nice application of the tensor virial theorem (1.46) (Nelson & Papaloizou 1993). The result is that the cloud will tend to collapse perpendicular to its long axis if the mass per unit length of cloud exceeds $2a^2/G$, where a is the isothermal sound speed.

10.3.1 *Jeans instability for a rotating system*

The Galaxy, and virtually everything within it, rotates. Unless it loses angular momentum by some mechanism, a rotating cloud will rotate faster as it collapses. Indeed, the centrifugal force will eventually balance self-gravity so that the cloud can no longer collapse perpendicularly to the rotation axis. (It can still collapse along the rotation axis.) We consider this scenario further in this section.

If the cloud is uniformly rotating, as we shall assume, it is convenient to work in a frame rotating with the cloud. In that frame the velocity of the initially non-collapsing cloud is zero. The equation of motion in the rotating frame is given by (3.3).

Equation (3.3) can be perturbed in the same manner as the perturbation analysis of Section 2.3. In equilibrium ($\boldsymbol{u} = 0$ in the rotating frame):

$$-\frac{1}{\rho_0}\nabla p_0 \; - \; \nabla\psi_0 \; - \; \Omega\times(\Omega\times r) \; = \; 0 \,, \tag{10.13}$$

and the linearized equation for small perturbations about equilibrium is

$$\frac{\partial \boldsymbol{u}}{\partial t} \; = \; -\frac{1}{\rho_0}\nabla p' \; + \; \frac{\rho'}{\rho_0^2}\nabla p_0 \; - \; \nabla\psi' \; - \; 2\Omega\times\boldsymbol{u} \,. \tag{10.14}$$

Once again we assume a uniform medium, so $\nabla p_0 = 0$, and suppose that fluctuations are isothermal: this yields the three linearized equations

$$\frac{\partial \boldsymbol{u}}{\partial t} \; = \; -\frac{a^2}{\rho_0}\nabla\rho' \; - \; \nabla\psi' \; - \; 2\Omega\times\boldsymbol{u} \,,$$

$$\nabla^2\psi' \; = \; 4\pi G\rho' \,, \tag{10.15}$$

$$\frac{\partial \rho'}{\partial t} \; = \; -\rho_0\nabla\cdot\boldsymbol{u} \,.$$

As before, we seek a solution with all perturbations proportional to $\exp(i\boldsymbol{k}\cdot\boldsymbol{x} - i\omega t)$. Setting up a coordinate system (x, y, z), we can without loss of generality take the z-axis along the $\boldsymbol{k}$ vector and then choose the other two axes so that Ω has no x-component:

$$\boldsymbol{k} = (0, 0, k), \qquad \boldsymbol{\Omega} = (0, \Omega \sin \theta, \Omega \cos \theta),$$

and we write $\boldsymbol{u} = (u, v, w)$. θ is the angle between the two vectors $\boldsymbol{k}$ and $\boldsymbol{\Omega}$. Then the components of the linearized perturbed equations can be written out in full as

$$\begin{aligned}
-i\omega u &= -2w\Omega \sin \theta + 2v\Omega \cos \theta \\
-i\omega v &= -2u\Omega \cos \theta \\
-i\omega w &= \frac{-ika^2}{\rho_0} \rho' - ik\psi' + 2u\Omega \sin \theta \\
-k^2 \psi' &= 4\pi G \rho'.
\end{aligned} \qquad (10.16)$$

These algebraic equations can be written in matrix form as

$$\begin{pmatrix}
i\omega & -2\Omega \cos \theta & 2\Omega \sin \theta & 0 & 0 \\
2\Omega \cos \theta & i\omega & 0 & 0 & 0 \\
-2\Omega \sin \theta & 0 & i\omega & ika^2/\rho_0 & ik \\
0 & 0 & 0 & -4\pi G & -k^2 \\
0 & 0 & ik\rho_0 & i\omega & 0
\end{pmatrix}
\begin{pmatrix}
u \\ v \\ w \\ \rho' \\ \psi'
\end{pmatrix} = \boldsymbol{0}.$$

$$(10.17)$$

If the matrix were invertible, we would have the trivial solution that $\boldsymbol{u} = \boldsymbol{0}$ and $\rho' = \psi' = 0$. Thus for a nontrivial solution, the determinant of the matrix must be zero. After some algebra one finds that the determinant is

$$k^2 \left[\omega^4 - (4\Omega^2 + \sigma^2)\omega^2 + 4\Omega^2 \sigma^2 \cos^2 \theta \right], \qquad (10.18)$$

where

$$\sigma^2 = a^2 k^2 - 4\pi G \rho_0. \qquad (10.19)$$

(Note that, in spite of the notation, σ^2 need not be positive.) Thus for zero determinant we deduce the dispersion relation that ω and $\boldsymbol{k}$ must satisfy:

$$\omega^4 - (4\Omega^2 + \sigma^2)\omega^2 + 4\Omega^2 \sigma^2 \cos^2 \theta = 0. \qquad (10.20)$$

For stability, both roots of this quadratic (in ω^2) must be positive. Now $x^2 - bx + c = 0$ has two positive roots if and only if $b > 0$ and $c > 0$. So the condition for *instability* is $b < 0$ or $c < 0$, i.e.

$$4\Omega^2 + \sigma^2 < 0 \qquad \text{or} \qquad \Omega^2 \sigma^2 \cos^2 \theta < 0.$$

Thus the cloud is unstable to growing perturbations if $\sigma^2 < 0$, that is, if $a^2 k^2 < 4\pi G\rho$.

So rotation does *not* change the Jeans instability criterion. However, it does change the rate at which density fluctuations grow.

In the non-rotating case $\omega^2 = \sigma^2$, so for $\sigma^2 < 0$ the fluctuations grow as $\exp(\eta t)$ where $\eta = -i\omega = |-\sigma^2|^{1/2}$.

In the rotating case

$$\omega^2 = \frac{1}{2}(4\Omega^2 + \sigma^2) \pm \frac{1}{2}\left[(4\Omega^2 + \sigma^2)^2 - 16\Omega^2\sigma^2\cos^2\theta\right]^{1/2} . \quad (10.21)$$

Consider for example $|\sigma^2| \gg \Omega^2$ and $\sigma^2 < 0$. Then (using the binomial theorem and neglecting terms of order Ω^4/σ^4)

$$\omega^2 \simeq \frac{1}{2}\left[(4\Omega^2 + \sigma^2) \mp (4\Omega^2 + \sigma^2 - 8\Omega^2\cos^2\theta)\right] , \quad (10.22)$$

i.e.

$$\omega^2 = \sigma^2 + 4\Omega^2\sin^2\theta \quad \text{or} \quad \omega^2 = 4\Omega^2\cos^2\theta .$$

The former is of course the unstable solution, and corresponds to growth rate

$$\eta = -i\omega = \left|(-\sigma^2) - 4\Omega^2\sin^2\theta\right|^{1/2}$$

which is *smaller* that the growth rate in the non-rotating case (note that σ^2 is negative). The growth rate is greatest for $\sin\theta = 0$, i.e. for $\boldsymbol{k}$ parallel to Ω, and smallest for $\boldsymbol{k}$ perpendicular to Ω. Thus the cloud tends to flatten into a disk. This could lead ultimately to the formation of a disk, which is indeed observed, and of planetary systems such as our own solar system.

10.4 Ambipolar Diffusion

The Galaxy rotates and has a weak magnetic field. We have just seen that the rotation modifies the dispersion relation for growing density perturbations and hinders collapse perpendicular to the rotation axis and so tends to lead to a flattened disk structure. To attain stellar densities however, the cloud must collapse also perpendicular to the rotation axis, and as it does so it spins up if it conserves its angular momentum. Likewise, if the magnetic field is frozen into the material as in ideal MHD (Chapter 5) then the field strength will also be magnified in inverse proportion to the reduction in cross-sectional area perpendicular to the field lines. Simple order-of-magnitude estimates suggest that this would lead to much stronger magnetic fields than are observed in young stars, and also that the proto-star that conserved its angular momentum would reach break-up velocity before it had collapsed sufficiently to attain stellar density.

A resolution of the magnetic problem is provided by *ambipolar diffu-sion*. The cloud is composed of both charged and neutral particles. The charged component is tied strongly to the magnetic field lines, but the neu-tral component is not directly tied to it, only indirectly through collisions with the charged particules. Collapse is therefore only partially inhibited by the magnetic field: while the neutrals collapse, there is some separating out of the charged and neutral species, with the charged particles and the magnetic field lagging behind in the collapse. Thus the collapsed core of the cloud will not have conserved the whole of its flux. The magnetic field can also help with the angular momentum problem. Provided the proto-star remains magnetically linked to surrounding matter, Alfvén waves can propagate along field lines and take angular momentum from the protostar and deposit it in the surrounding medium. Thus the collapsing core of the cloud does not conserve the whole of its angular momentum.

10.5   Fragmentation

For a typical interstellar HI region, the number density of hydrogen atoms is $n_H = 10^7 \, \mathrm{m^{-3}}$, so the density $m_H n$ is about $2 \times 10^{-20} \, \mathrm{kg \, m^{-3}}$. The temperature is about $100 \, \mathrm{K}$. Hence, using $a^2 = \mathcal{R}T/\mu$ with mean molecular weight $\mu \approx 1$, one gets that $\lambda_J \approx 10^{18} \, \mathrm{m}$ (about $30 \, \mathrm{pc}$) and $M_J \approx 3 \times 10^{34} \, \mathrm{kg}$ $\approx 10^4$ solar masses. But this is very much greater than the mass of the most massive known stars. Hence it cannot simply be that a cloud exceeding its Jeans mass collapses under its own self-gravity to form a single star.

This was recognised by Hoyle (1953). A possible resolution to the prob-lem is a hierarchical process of fragmentation. If we suppose that the col-lapse continues isothermally (i.e. a^2 remains constant) then $\lambda_J \propto \rho^{-1/2}$ and similarly $M_J \propto \rho^{-1/2}$. Hence, as the cloud collapses and the density grows, the Jeans mass decreases. One can therefore envisage that subcondensa-tions form in the cloud as smaller and smaller masses start to collapse in upon themselves. This picture is known as fragmentation. However, frag-mentation must not continue indefinitely, as we wish eventually to form stars with observed stellar masses. In the later stages of collapse, the cloud presumably becomes too opaque for radiation to smooth out temperature fluctuations, so it is reasonable to suppose that isothermal collapse is no longer a good assumption. If we assume that we enter the opposite regime, of adiabatic collapse, then

$$\frac{1}{p}\frac{Dp}{Dt} = \frac{\gamma}{\rho}\frac{D\rho}{Dt} \qquad (10.23)$$

so $p \propto \rho^\gamma$. Instead of the isothermal sound speed we use the adiabatic sound speed c: $c^2 = \gamma p/\rho \propto \rho^{\gamma-1}$. From (10.10) and (10.11) we then have $\lambda_J \propto \rho^{(\gamma-2)/2}$ and $M_J \propto \rho^{(3\gamma/2-2)}$. For example, for $\gamma = 5/3$, $M_J \propto \rho^{1/2}$. Hence the Jeans mass is no longer decreasing with increasing cloud density, and this can halt the process fragmentation.

A problem with the fragmentation picture as propounded above is that dispersion relation (10.9) implies that ω^2 is most negative for the smallest values of k. Thus, the cloud collapses fastest at the largest scales. Although the cloud becomes unstable on smaller scales as the density increases, these smaller-scale perturbations would get overwhelmed by the faster overall collapse of the cloud.

10.6 Some Comments on Star Formation

Real star formation is more complicated than mere consideration of the Jeans instability. Much observational and computational work has helped form a more detailed picture of how stars form. The above ideas nonetheless provide a useful framework for interpreting the detailed results.

Dense, cold interstellar clouds of molecular hydrogen are preferred sites for star formation: with their lower temperatures and higher densities, they have the advantage of having lower Jeans masses. How such molecular clouds form is itself not fully understood: density waves propagating through the galaxy and local supernova explosions may play a role in assembling them from less dense ionized or neutral atomic hydrogen. In our discussion above we have supposed that the pressure support inside the cloud comes from gas pressure, but interstellar clouds can be highly turbulent and turbulent pressure support is likely to be important in controlling star formation in many if not all situations. That being the case, a possible scenario for the formation in particular of clusters of stars in giant molecular clouds (e.g. Lada & Lada 2003) is that colliding flows inside a cloud cause shocks and the dissipation of energy and reduction of turbulent pressure locally leads to the creation of overdense regions that proceed to collapse rapidly and condense out of the surrounding turbulent medium. Massive collapsed cores may eventually form clusters of stars, through fragmentation processes (see previous section). Shocks within the turbulent flows can also contribute to fragmentation. Embryonic protostar cores also gain mass by accretion through infall of surrounding material (e.g. Clarke *et al.* 2000). In the heart of forming stellar clusters, it is possible and even likely that massive stars can form by collisions and mergers between protostellar cores.

The formation of isolated, low-mass stars may be quite different. Indeed dwarf molecular clouds are observed that have masses much greater than their Jeans mass, which may only be a few solar masses (Shu *et al.* 1987), and yet they are not collapsing on a dynamical timescale. The rate at which protostars condense out in such systems may be regulated by magnetic field, the collapse proceeding in a quasi-equilibrium as ambipolar diffusion gradually allows infalling material to slip relative to the magnetic field. If massive stars form in a stellar nursery, it is thought that they may switch off further star formation. Once they start to burn hydrogen, their high luminosities lead to substantial radiative pressure and winds blow away the raw material for any further star formation.

A final point is that the formation of a pair of collimated jets is a common observational feature of young stellar objects (YSOs) and indeed all young stars may go through an outflow phase. What forms these directed outflows is uncertain. A possible scenario is that the stars produce an isotropic stellar wind but that this can escape more easily along the polar axis than in the equatorial plane of the star, where rotational flattening will have formed a disk of material. Hence the wind emerges as collimated jets along the rotation axis of the star.

Chapter 11

Radial Oscillations of Stars

In this chapter we consider oscillations of a star about its spherically symmetric equilibrium state, with the oscillatory motion being purely radial. This is relevant to a number of classical variable stars, e.g. Cepheids. The position of various types of classical variable stars are shown on the H-R diagram in Fig. 11.1, including the Cepheids, which are located in the classical instability strip. Classes of variable stars are often named after the first well-studied example of their type: the Cepheid class is named after the progenitor star, δ Cephei.

Other classes shown in Fig. 11.1 are the giant Myra variables and (beneath them in the figure) the irregular variables; RR Lyrae stars; on or near the main sequence (in order of decreasing temperature) the β Cephei stars, slowly pulsating B stars (SPBs), δ Scuti stars, rapidly oscillating chemically peculiar A stars (roAp stars), γ Doradus stars, and solar-like stars which, like the Sun, undergo stochasically excited small-amplitude oscillations; on the helium core-burning horizontal branch the sub-dwarf variable B stars (sdBs); and along the white-dwarf cooling track the planetary nebula nuclei variables (PNNVs) and three classes of white-dwarf variables, DOV, DBV and DAV stars. Most of these actually oscillate in nonradial modes, the theory of which is developed in Chapter 12.

A readable account of the theory of radial stellar pulsations may be found in Cox (1980).

11.1 Linear Adiabatic Wave Equation for Radial Oscillations

We assume that the amplitude of the oscillations is small, so that linear perturbation theory will suffice, and in the present section we suppose that

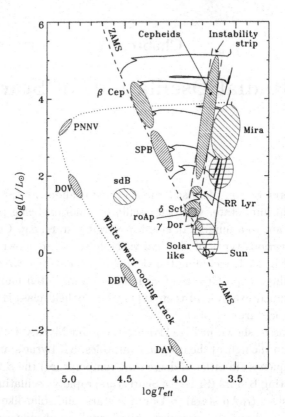

Fig. 11.1   Hertzsprung-Russell diagram showing locations of some classes of stellar pulsators, as a function of stellar luminosity L and effective temperature T_{eff}. The location of the zero-age main sequence (ZAMS) is shown as a dashed curve. Evolution tracks of selected stars of different mass are shown as solid curves. The cooling track along which planetary nebulae and white dwarfs evolve is shown as a dotted curve. The classical instability strip lies between the pair of long-dashed lines. The hatching indicates the type of oscillations exhibited in the stars (see Chapter 12): horizontal hatching indicates stochastically excited oscillations, hatching sloping from top left to bottom right indicates p-mode oscillations, and hatching sloping from bottom left to top right indicates g-mode oscillations. (Courtesy J. Christensen-Dalsgaard.)

the period of the oscillations is sufficiently short that no heat is exchanged between neighbouring fluid elements, i.e. the oscillations are adiabatic. Then the linearized perturbed equation of motion is (2.41):

$$\frac{\partial^2 \boldsymbol{u}}{\partial t^2} = \frac{1}{\rho_0} \nabla \left[\gamma p_0 \nabla \cdot \boldsymbol{u} + \boldsymbol{u} \cdot \nabla p_0 \right] - \frac{1}{\rho_0^2} \nabla \cdot (\rho_0 \boldsymbol{u}) \nabla p_0 - \nabla \left(\frac{\partial \psi'}{\partial t} \right) ; \tag{11.1}$$

while differentiating the perturbed Poisson equation (2.34) and using the continuity equation (2.33) gives

$$\nabla^2 \frac{\partial \psi'}{\partial t} = -4\pi G \nabla \cdot (\rho_0 \boldsymbol{u}) . \tag{11.2}$$

Specializing to the case $\boldsymbol{u} = u_r \boldsymbol{e}_r$, and with all quantities independent of spherical polar coordinates θ and ϕ, Eq. (11.2) becomes

$$\frac{1}{r^2} \frac{\partial}{\partial r} \left[r^2 \frac{\partial}{\partial r} \left(\frac{\partial \psi'}{\partial t} \right) \right] = \frac{-4\pi G}{r^2} \frac{\partial}{\partial r} (r^2 \rho_0 u_r)$$

which can be integrated to give

$$r^2 \frac{\partial}{\partial r} \left(\frac{\partial \psi'}{\partial t} \right) = -4\pi G r^2 \rho_0 u_r + f(t)$$

where $f(t)$ is a "constant" of integration. Evaluating this expression at $r = 0$ shows that $f(t) = 0$ for all t, and hence we have that

$$\nabla \left(\frac{\partial \psi'}{\partial t} \right) = -4\pi G \rho_0 u_r \boldsymbol{e}_r . \tag{11.3}$$

Substituting (11.3) into the radial component of (11.1) gives

$$\frac{\partial^2 u_r}{\partial t^2} = \frac{1}{\rho_0} \frac{\partial}{\partial r} \left[\frac{\gamma p_0}{r^2} \frac{\partial}{\partial r} (r^2 u_r) + u_r \frac{dp_0}{dr} \right]$$
$$- \frac{1}{\rho_0^2 r^2} \frac{\partial}{\partial r} (r^2 \rho_0 u_r) \frac{dp_0}{dr} + 4\pi G \rho_0 u_r . \tag{11.4}$$

Let us define a new variable $\xi = u_r/r$. Then Eq. (11.4) becomes

$$\frac{\partial^2 \xi}{\partial t^2} = \frac{1}{\rho_0 r} \frac{\partial}{\partial r} \left[\frac{\gamma p_0}{r^2} \frac{\partial}{\partial r} (r^3 \xi) + r\xi \frac{dp_0}{dr} \right]$$
$$- \frac{1}{\rho_0^2 r^3} \frac{\partial}{\partial r} (r^3 \rho_0 \xi) \frac{dp_0}{dr} + 4\pi G \rho_0 \xi , \tag{11.5}$$

which after some re-arranging can be written as

$$\rho_0 r \frac{\partial^2 \xi}{\partial t^2} = \frac{1}{r^3} \frac{\partial}{\partial r} \left(\gamma p_0 r^4 \frac{\partial \xi}{\partial r} \right) + \frac{d}{dr} \left[(3\gamma - 4) p_0 \right] \xi . \tag{11.6}$$

Since the coefficients in Eq. (11.6) are independent of time, we may seek solutions with sinusoidal time dependence. We write

$$\xi(r, t) = \xi(r) e^{-i\omega t} ,$$

where for reasons of economy of notation we have used the same symbol ξ to denote both the full variable with time dependence and its time-independent amplitude. Then Eq. (11.6) becomes

$$\frac{d}{dr}\left(\gamma p_0 r^4 \frac{d\xi}{dr}\right) + r^3 \frac{d}{dr}\left[(3\gamma - 4)p_0\right]\xi + \rho_0 r^4 \omega^2 \xi = 0 . \tag{11.7}$$

We refer to this as the linear adiabatic wave equation for radial oscillations.

Henceforth we shall omit the zero subscripts on equilibrium quantities, except where confusion might arise.

11.1.1   *Boundary conditions*

Equation (11.7) is an ordinary differential equation for the amplitude function $\xi(r)$. In order to solve it we require boundary conditions at the centre of the star ($r = 0$) and at its surface ($r = R$). For simplicity, in deriving the boundary conditions we shall assume that γ is a constant.

The boundary condition at $r = 0$ is the requirement that the solution be regular, not divergent, at the origin. We postulate that near $r = 0$

$$\xi = r^\alpha \sum_{n=0}^{\infty} a_n r^n \qquad (a_0 \neq 0) . \tag{11.8}$$

Also we write power series expansions of p and ρ in the neighbourhood of $r = 0$:

$$p = p_c\left(1 - p_2 r^2 + \dots\right) , \qquad \rho = \rho_c\left(1 - \dots\right) . \tag{11.9}$$

Note that there is no term linear in r in the expansion of p: this is because $m \simeq (4\pi/3)\rho_c r^3$ near the origin; hence the right-hand side of the hydrostatic equation (2.11) vanishes as $r \to 0$ and thus so too must dp/dr. Substituting expansions (11.8) and (11.9) into Eq. (11.7) yields

$$\gamma p_c \alpha(\alpha+3)a_0 r^{\alpha+2} + \gamma p_c(\alpha+1)(\alpha+4)a_1 r^{\alpha+3} + O(r^{\alpha+4}) = 0 . \tag{11.10}$$

Each coefficient in this equation must separately vanish. Hence from the $O(r^{\alpha+2})$ term we have that $\alpha = 0$ or $\alpha = -3$ (because $a_0 \neq 0$). The requirement that the solution be regular rules out $\alpha = -3$: hence we must have $\alpha = 0$. Having determined α, the coefficient of $r^{\alpha+3}$ determines a_1: $a_1 = 0$. This implies that $d\xi/dr$ vanishes at the origin. Hence a practical boundary condition, that picks out the regular solution rather than the divergent one, is

$$\frac{d\xi}{dr} = 0 \quad \text{at} \quad r = 0 . \tag{11.11}$$

The simplest reasonable approximation to make at the surface is that $p = 0$ at $r = R$. This is not precisely true for a real star, but for the Sun for example the ratio of the surface (photospheric) pressure to the central pressure is 10^{-23}, so it is a reasonable approximation. Likewise, p/ρ, which is essentially temperature, is very small at the surface compared with its value at the centre of the star — the ratio is $\simeq 4 \times 10^{-4}$ for the Sun — and hence we shall assume that $p/\rho = 0$ at $r = R$. Expanding Eq. (11.7) gives

$$\gamma p r^4 \frac{d^2\xi}{dr^2} + \left(4\gamma p r^3 - \gamma G m \rho r^2\right) \frac{d\xi}{dr} - (3\gamma - 4)G m \rho r \xi$$
$$+ \rho \omega^2 r^4 \xi = 0$$

which, after some re-arrangement, yields

$$\left(\frac{\gamma p r^3}{Gm\rho}\right) \left[r\frac{d^2\xi}{dr^2} + 4\frac{d\xi}{dr}\right] = \gamma r^2 \left[\frac{d\xi}{dr} - \left\{\frac{\omega^2 r^2}{\gamma Gm} - \frac{(3\gamma - 4)}{\gamma r}\right\}\xi\right] . \tag{11.12}$$

This equation is true everywhere and so, in particular, at the surface. At $r = R$, p/ρ vanishes, and so both sides of Eq. (11.12) must be zero there. Hence

$$\frac{d\xi}{dr} = \frac{1}{\gamma R} \left[\frac{\omega^2 R^3}{GM} - (3\gamma - 4)\right] \xi \quad \text{at } r = R \tag{11.13}$$

where M is the value of m at $r = R$ (so, essentially, the total mass of the star). This provides the second boundary condition. As before, this boundary condition serves to pick out the regular solution rather than the divergent one.

11.1.2 Eigenvalue nature of the problem

The second-order ordinary differential Eq. (11.7) and boundary conditions (11.11) and (11.13) are linear and homogeneous in ξ. Solving the differential equation gives two constants of integration, but the amplitude of the solution will be arbitrary so one constant cannot be determined. The other is determined by one of the boundary conditions. The second boundary condition can only then be satisfied for certain values of ω^2: such values

are called eigenvalues. These eigenvalues give the resonant frequencies of
the radial oscillations of a star.

11.1.3 *Self-adjointness of the problem*

Equation (11.7) can be written as

$$\mathcal{L}\xi + \rho\omega^2 r^4 \xi = 0 , \tag{11.14}$$

where

$$\mathcal{L}\xi \equiv \frac{d}{dr}\left(\gamma p r^4 \frac{d\xi}{dr}\right) + r^3 \frac{d}{dr}\left[(3\gamma - 4)p\right]\xi . \tag{11.15}$$

Suppose ξ and η are any two functions that both satisfy the boundary
conditions. The operator $\mathcal{L}$ (with the associated boundary conditions) is
said to be *self-adjoint* if

$$\int_0^R \eta^* \mathcal{L}\xi \, dr = \int_0^R \xi \left(\mathcal{L}\eta\right)^* dr , \tag{11.16}$$

where the star denotes the complex conjugate. Using definition (11.15), it
is straightforward to show that

$$\int_0^R \eta^* \mathcal{L}\xi - \xi \left(\mathcal{L}\eta\right)^* dr = \int_0^R \eta^* \frac{d}{dr}\left(\gamma p r^4 \frac{d\xi}{dr}\right) - \xi \frac{d}{dr}\left(\gamma p r^4 \frac{d\eta^*}{dr}\right) dr$$

$$= \left[\eta^* \gamma p r^4 \frac{d\xi}{dr} - \xi \gamma p r^4 \frac{d\eta^*}{dr}\right]_0^R = 0$$

since $d\xi/dr = 0$ at $r = 0$, and $p = 0$ at $r = R$. Hence $\mathcal{L}$ *is* self-adjoint in
our case.

Various nice properties follow (see Appendix C). In particular, the eigenvalues ω^2 are *real*, and any two eigenfunctions ξ_1, ξ_2 corresponding to distinct eigenvalues ω_1^2, ω_2^2 are orthogonal in the sense that

$$\int_0^R \rho r^4 \xi_2^* \xi_1 \, dr = 0 . \tag{11.17}$$

We now define

$$F[\xi] = \frac{-\int_0^R \xi^* \mathcal{L}\xi \, dr}{\int_0^R \rho r^4 \xi^* \xi \, dr} . \tag{11.18}$$

Clearly, if ξ is an eigenfunction, then by Eq. (11.14) $F[\xi]$ is equal to the corresponding eigenvalue ω^2. But further, if ξ is any function (not necessarily an eigenfunction) satisfying the boundary conditions (essentially that $d\xi/dr = 0$ at $r = 0$ and that ξ is regular at $r = R$), then

$$F[\xi] \geq \omega_0^2 , \tag{11.19}$$

where ω_0^2 is the smallest eigenvalue. In our case, ω_0 is the frequency of the *fundamental mode*: thus for any "trial function" ξ, $F[\xi]$ provides an upper bound on the fundamental frequency.

In our pulsation problem,

$$
\begin{aligned}
-\int_0^R \xi^* \mathcal{L}\xi \, dr &= -\int_0^R \xi^* \frac{d}{dr}\left(\gamma p r^4 \frac{d\xi}{dr}\right) dr - \int_0^R \xi^* r^3 \frac{d}{dr}[(3\gamma - 4)p]\,\xi \, dr \\
&= \int_0^R \gamma p r^4 \left|\frac{d\xi}{dr}\right|^2 dr - \int_0^R r^3 \frac{d}{dr}[(3\gamma - 4)p]\,|\xi|^2 \, dr
\end{aligned}
$$

integrating by parts (the surface term vanishes using the boundary conditions). Hence

$$F[\omega] = \frac{\int_0^R \gamma p r^4 \left|\frac{d\xi}{dr}\right|^2 dr - \int_0^R r^3 \frac{d}{dr}[(3\gamma - 4)p]\,|\xi|^2 \, dr}{\int_0^R \rho r^4 |\xi|^2 \, dr} . \tag{11.20}$$

The simplest suitable trial function is $\xi = \text{constant}$. Then

$$F[\xi] = \frac{0 - \int_0^R r^3 \frac{d}{dr}[(3\gamma - 4)p]\, dr}{\int_0^R \rho r^4 \, dr} = \frac{\int_0^R 3r^2(3\gamma - 4)p \, dr}{\int_0^R \rho r^4 \, dr} ,$$

integrating by parts. Hence

$$\omega_0^2 \leq \frac{\int_0^R 3(3\gamma - 4)p r^2 \, dr}{\int_0^R \rho r^4 \, dr} \tag{11.21}$$

gives an upper bound on the frequency of the fundamental mode.

Note that if γ were everwhere less than $4/3$, then the fundamental frequency would be imaginary, corresponding to an exponentially growing perturbation. Thus one cannot have a stable stellar model with $\gamma < 4/3$ everywhere.

11.1.4   *A lower bound on the fundamental frequency*

If ξ is an eigenfunction, then the right-hand side of Eq. (11.20) is equal to the corresponding eigenvalue ω^2. Noting that the first term in the numerator must be non-negative, and assuming that γ can be treated as a constant, we thus obtain

$$\omega^2 \geq \frac{-\int_0^R r^3 \frac{d}{dr}\left[(3\gamma-4)p\right]|\xi|^2\,dr}{\int_0^R \rho r^4 |\xi|^2\,dr} \geq (3\gamma-4)\frac{\int_0^R Gm\rho r|\xi|^2\,dr}{\int_0^R \rho r^4 |\xi|^2\,dr}$$

$$\geq (3\gamma-4)\frac{4\pi}{3}G\bar\rho(R)\frac{\int_0^R \frac{\bar\rho(r)}{\bar\rho(R)}\rho r^4 |\xi|^2\,dr}{\int_0^R \rho r^4 |\xi|^2\,dr} . \qquad (11.22)$$

In the final step we have written $m(r) = (4\pi/3)r^3\bar\rho(r)$, i.e. $\bar\rho(r)$ is the mean density of the material within the sphere of radius r. In most stars, $\bar\rho(r)$ is a decreasing function of r, so $\bar\rho(r)/\bar\rho(R) \geq 1$ everywhere. In that case, Eq. (11.22) implies

$$\omega^2 \geq (3\gamma-4)\frac{4\pi}{3}G\bar\rho(R) ,$$

i.e.

$$\omega^2 \geq (3\gamma-4)\frac{GM}{R^3} . \qquad (11.23)$$

This provides a *lower* bound on the fundamental frequency. Put another way, it provides an upper bound on the period $P_0 \equiv 2\pi/\omega_0$ of the fundamental mode.

Equation (11.23) implies that if $\gamma > 4/3$ then all radial modes are stable ($\omega^2 > 0$) in the linear, adiabatic approximation.

11.1.5   *Homology scaling for the fundamental frequency of stars*

Suppose that p, ρ and γ vary homologously from one star to another: i.e. making p, ρ and r dimensionless using G, M and R, the dimensionless functions $\tilde p(x)$, $\tilde\rho(x)$ and $\gamma(x)$ are the same for each star under consideration (x being r/R). Then Eq. (11.7) can be rewritten in the dimensionless form

$$\frac{d}{dx}\left[\gamma\tilde p x^4 \frac{d\xi}{dx}\right] + x^3\frac{d}{dx}\left[(3\gamma-4)\tilde p\right]\xi + \tilde\rho\tilde\omega^2 x^4\xi = 0 , \qquad (11.24)$$

where

$$\omega^2 = \left(\frac{GM}{R^3}\right)\tilde{\omega}^2 . \tag{11.25}$$

The eigensolutions of (11.24), and in particular the dimensionless eigenvalues $\tilde{\omega}^2$, are thus the same for all the stars, and so the dimensional frequencies ω scale from one star to another as

$$\left(\frac{GM}{R^3}\right)^{1/2} \propto (\rho_{\text{mean}})^{1/2} .$$

Hence, in particular, the fundamental period scales as $(\rho_{\text{mean}})^{-1/2}$.

Even though different stars may be not at all homologous to one another, this scaling gives a reasonable estimate of their fundamental periods. The fundamental period of the Sun ($\rho_{\text{mean}} \sim 1400 \, \text{kg m}^{-3}$) is about one hour. For a white dwarf ($\rho_{\text{mean}} \sim 10^9 \, \text{kg m}^{-3}$) the fundamental period is a few seconds; while for a tenuous red supergiant ($\rho_{\text{mean}} \sim 10^{-6} \, \text{kg m}^{-3}$) it is of the order of 10^3 days.

11.2 Non-adiabatic Radial Oscillations

The theory of the previous section provides a good approximation to the frequencies of the radial oscillations of stars. However, the adiabatic theory cannot address issues relating to the excitation and damping of oscillations, because in the adiabatic theory there is no net transfer of energy to or from the oscillation over a complete period. To describe non-adiabatic effects, we need to perturb the energy Eq. (2.4). Actually, it is more convenient to use the energy equation in the first of forms (1.28) (recalling that this refers to a fixed mass of fluid):

$$\frac{Dp}{Dt} - \frac{\gamma p_0}{\rho_0}\frac{D\rho}{Dt} = \rho_0(\gamma_3 - 1)\left(\epsilon_0 + \delta\epsilon - \frac{\partial(L_0 + \delta L)}{\partial m}\right) . \tag{11.26}$$

The zero subscript has temporarily been restored, for clarity, and we have used the fact that, for a spherically symmetric star, $\nabla \cdot \boldsymbol{F} = \rho \partial L/\partial m$, L being luminosity. Now, in equilibrium, $\epsilon_0 = \partial L_0/\partial m$; also, in linear perturbation theory,

$$\frac{Dp}{Dt} = \frac{D\delta p}{Dt} = \frac{\partial p'}{\partial t} + \boldsymbol{u} \cdot \nabla p \tag{11.27}$$

(and similarly for any other perturbed quantity). Hence, Eq. (11.26) implies that

$$\frac{\partial p'}{\partial t} + \boldsymbol{u}\cdot\nabla p = \frac{\gamma p}{\rho}\left(\frac{\partial \rho'}{\partial t} + \boldsymbol{u}\cdot\nabla\rho\right) + \rho(\gamma_3 - 1)\left(\delta\epsilon - \frac{\partial\delta L}{\partial m}\right). \quad (11.28)$$

Using this equation in place of Eq. (2.40), Eq. (11.7) becomes, in non-adiabatic theory,

$$\frac{\mathrm{d}}{\mathrm{d}r}\left(\gamma p r^4 \frac{\mathrm{d}\xi}{\mathrm{d}r}\right) + r^3 \frac{\mathrm{d}}{\mathrm{d}r}\left[(3\gamma - 4)p\right]\xi + \rho r^4 \omega^2 \xi$$
$$= r^3 \frac{\mathrm{d}}{\mathrm{d}r}\left[(\gamma_3 - 1)\rho\left(\delta\epsilon - \frac{\mathrm{d}\delta L}{\mathrm{d}m}\right)\right]. \quad (11.29)$$

The same factor $\exp(-i\omega t)$ has been taken out of $\delta\epsilon$ and δL as out of ξ, so that all three are functions of r only.

Although we shall not use it, we note here that one further equation is necessary (in addition to an equation of state and expressions for the energy generation rate and opacity κ) to get a complete set of equations describing non-adiabatic oscillations: this is an equation describing the energy transport. If, for example, energy is transported wholly by diffusive radiative transfer, then the appropriate equation would be

$$\frac{\mathrm{d}}{\mathrm{d}r}\left(\frac{\delta T}{T}\right) = \frac{\mathrm{d}\ln T}{\mathrm{d}r}\left[\frac{\delta L}{L} - 4\frac{\delta r}{r} + \frac{\delta\kappa}{\kappa} - 4\frac{\delta T}{T}\right], \quad (11.30)$$

which we have obtained by perturbing Eq. (2.26).

In adiabatic theory, the oscillation period is assumed to be too short for any heat to be transported. At the oppositive extreme, one may consider oscillations that take place so slowly that the star always has time to adjust and remain in hydrostatic equilibrium. Such oscillations are called thermal modes: in this case, it is appropriate to drop the last term on the left-hand side of Eq. (11.29), since this came from the acceleration term in the equation of motion. Thus, thermal modes can be investigated by solving the modified Eq. (11.29), together with an equation such as (11.30). The relevant timescale for such motions is the thermal one: for the Sun, the global thermal timescale (the Kelvin-Helmholtz timescale) is of the order of 10^7 years, about twelve orders of magnitude longer than the dynamical timescale. We shall not consider thermal modes further in these notes. For more details see Hansen (1987) (but note that there is a factor ω missing from the third term of their Eq. [1.4]).

Multiplying Eq. (11.29) by ξ^* and integrating gives

$$\int_0^R -\gamma p r^4 \left|\frac{d\xi}{dr}\right|^2 + r^3 \frac{d}{dr} \left[(3\gamma - 4)p\right] |\xi|^2 \, dr + \omega^2 \int_0^R \rho r^4 |\xi|^2$$

$$= \int_0^R r^3 \xi^* \frac{d}{dr} \left[(\gamma_3 - 1)\rho \left(\delta\epsilon - \frac{d\delta L}{dm}\right)\right] dr$$

$$= \int_0^R -\frac{d}{dr}(r^3 \xi^*)(\gamma_3 - 1)\rho \left(\delta\epsilon - \frac{d\delta L}{dm}\right) dr$$

$$= i\omega^* \int_0^R \delta\rho^*(\gamma_3 - 1) \left(\delta\epsilon - \frac{d\delta L}{dm}\right) r^2 \, dr \ . \qquad (11.31)$$

Here we have used the mass conservation equation

$$\frac{D\rho}{Dt} = -\rho \nabla \cdot \boldsymbol{u} = -\frac{\rho}{r^2} \frac{d}{dr}(r^3 \xi)$$

so that

$$-i\omega\delta\rho = \frac{-\rho}{r^2} \frac{d}{dr}(r^3 \xi) \ .$$

Let

$$C = 4\pi \int_0^R \delta\rho^*(\gamma_3 - 1) \left(\delta\epsilon - \frac{d\delta L}{dm}\right) r^2 \, dr$$

$$I = 4\pi \int_0^R \rho r^4 |\xi|^2 \, dr$$

$$I\Sigma^2 = 4\pi \int_0^R \gamma p r^4 \left|\frac{d\xi}{dr}\right|^2 - r^3 \frac{d}{dr} \left[(3\gamma - 4)p\right] |\xi|^2 \, dr \ . \qquad (11.32)$$

Then (11.31) can be written as

$$I\left(\omega^2 - \Sigma^2\right) = i\omega^* C \ .$$

Thus

$$\omega = \pm\Sigma \left(1 + \frac{i\omega^* C}{I\Sigma^2}\right)^{1/2} \ .$$

If the oscillations are only weakly non-adiabatic (i.e. C is small), then to a first approximation $\omega = \pm\Sigma$. Assuming that Σ is real, this can then be substituted into the small term involving C, yielding

$$\omega \approx \pm\Sigma \left(1 \pm \frac{iC}{2I\Sigma}\right) = \pm\Sigma + \frac{iC}{2I} \qquad (11.33)$$

Hence

$$e^{-i\omega t} \approx e^{\mp i\Sigma t} e^{(C/2I)t} .$$

Thus the oscillations grow if $\mathrm{Re}(C) > 0$.

11.2.1 *Physical discussion of driving and damping*

Let us now consider the mechanical energy Eq. (1.16), with $\boldsymbol{f} = -\nabla\psi$. Integrating over a given mass of fluid

$$\int \frac{D}{Dt}\left(\frac{1}{2}u^2\right) \mathrm{d}m = -\int \rho^{-1}\boldsymbol{u}\cdot\nabla p \,\mathrm{d}m - \int \boldsymbol{u}\cdot\nabla\psi \,\mathrm{d}m$$

$$= -\int p\boldsymbol{u}\cdot\boldsymbol{n}\,\mathrm{d}S + \int \frac{p}{\rho}\nabla\cdot\boldsymbol{u}\,\mathrm{d}m - \int \boldsymbol{u}\cdot\nabla\psi\,\mathrm{d}m$$

after application of the divergence theorem (noting that a mass element $\mathrm{d}m$ is density times a volume element), where the surface integral is over the surface of the region occupied by the fluid mass. The surface term vanishes because $p = 0$ at $r = R$; so using the mass conservation equation, this gives

$$\frac{\mathrm{d}}{\mathrm{d}t}\int \frac{1}{2}u^2\,\mathrm{d}m = \int p\frac{D}{Dt}\left(\frac{1}{\rho}\right)\mathrm{d}m - \int \boldsymbol{u}\cdot\nabla\psi\,\mathrm{d}m . \qquad (11.34)$$

One can think of the integrals in Eq. (11.34) as summations over concentric mass shells within the star. Equation (11.34) then says that the rate of change of kinetic energy (in the pulsation) is equal to

$$\sum_{\text{all mass shells}} (p\,\mathrm{d}V \text{ per unit time})\,\mathrm{d}m \qquad (11.35)$$

minus a conservative gravitational term, where V is the thermodynamic variable, volume per unit mass. If we integrate over one cycle of the oscillation, the conservative gravitational term gives zero contribution, and we find that the increase in pulsational energy over one cycle is equal to

$$\sum_{\text{all mass shells}} \mathrm{d}m \times \oint p\,\mathrm{d}V .$$

Consider the two cases illustrated in Fig. 11.2. In case (i), $\oint p\,\mathrm{d}V > 0$: hence this tends to *drive* oscillations. In case (ii), $\oint p\,\mathrm{d}V < 0$ and so this tends to *damp* the oscillations. Now from Eq. (1.28),

$$\frac{1}{p}\frac{Dp}{Dt} = \frac{\gamma}{\rho}\frac{D\rho}{Dt} + (\gamma_3 - 1)\rho\left(\epsilon - \frac{1}{\rho}\nabla\cdot\boldsymbol{F}\right) . \qquad (11.36)$$

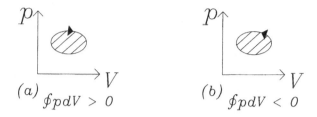

Fig. 11.2   The work done in one cycle, at some point in the star: (a) $\oint p\,dV$ is positive, which tends to drive oscillations; (b) $\oint p\,dV$ is negative, which tends to damp oscillations.

(We leave the reference to γ_3 explicit to distinguish it from other instances of γ in this chapter, which have been γ_1.)

In the driving case, at maximum compression (V minimum; $D\rho/Dt = 0$) p is still increasing and so, by Eq. (11.36), heat is being added. Thus, for oscillations to be driven, at maximum compression heat is being gained. Also, p lags behind ρ.

For oscillations to be damped, at maximum compression, heat is being lost. Also, ρ lags behind p.

One way in which oscillations can be driven is by the so-called κ-mechanism, so because in stellar structure theory the opacity to radiation is customarily denoted by κ (see Section 2.2). A power law approximation to the opacity is often reasonably good: $\kappa \propto \rho^\alpha T^{-\beta}$. For Kramers' opacity, for example, $\alpha = 1$ and $\beta = -7/2$. Then changes in opacity are related to changes in density and temperature as

$$\frac{\delta\kappa}{\kappa} = \alpha\frac{\delta\rho}{\rho} + \beta\frac{\delta T}{T} \ . \tag{11.37}$$

If the oscillations are only weakly non-adiabatic, we can use Eq. (1.27) to write

$$\delta T/T \simeq (\gamma_3 - 1)\delta\rho/\rho \ , \tag{11.38}$$

and so, combining these two equations,

$$\frac{\delta\kappa}{\kappa} \simeq [\alpha + \beta(\gamma_3 - 1)]\frac{\delta\rho}{\rho} \ . \tag{11.39}$$

Usually, when a star is compressed ($\delta\rho > 0$), $\delta\kappa$ is negative (because $\gamma_3 \approx 5/3$, and α and β take e.g. Kramers values) and so it is easier for heat to escape. In a region of partial ionization, however, the value of γ_3 is reduced and the logarithmic derivative of opacity with respect of temperature (β) in particular may also change. The term in square brackets in Eq. (11.39) can in these circumstances become positive, in which case the opacity will *increase* when the star is compressed. In that case, heat gets dammed up at maximum compression, causing oscillations to be driven. This was termed *overstability* by Eddington.

11.3 The Quasi-adiabatic Approximation

One can think of each mass shell within a star as a heat engine. That shell will be a driving region if it gains heat at maximum compression, and a damping region if it loses heat at maximum compression. (Note that, for adiabatic oscillations, heat is neither gained nor lost.) Some parts of the star will drive, others will damp. Whether an oscillation is driven or damped will be determined by which wins on balance.

More quantitatively, we look at

$$
C = 4\pi \int_0^R \delta\rho^*(\gamma_3 - 1)\left(\delta\epsilon - \frac{d\delta L}{dm}\right) r^2\, dr
$$

$$
= \int_0^M (\gamma_3 - 1)\left(\frac{\delta\rho}{\rho}\right)^*\left(\delta\epsilon - \frac{d\delta L}{dm}\right) dm .
\tag{11.40}
$$

Recall that oscillations are driven if $\mathrm{Re}(C) > 0$, damped if $\mathrm{Re}(C) < 0$. Driving regions are where the real part of the integrand of C is positive; damping regions are where the real part of the integrand is negative.

One should properly use the actual non-adiabatic eigenfunctions in this integral. What is often done is to use the adiabatic eigenfunctions, which are easier to compute. This is called the *quasi-adiabatic* approximation.

Consider the first term ($\delta\epsilon$) in the integrand. If $\epsilon \approx \epsilon_0\rho T^\eta$, where ϵ_0 and η are constants and η is positive, then

$$
\frac{\delta\epsilon}{\epsilon} \approx \frac{\delta\rho}{\rho} + \eta\frac{\delta T}{T} \approx [1 + \eta(\gamma_3 - 1)]\frac{\delta\rho}{\rho} ,
\tag{11.41}
$$

using the adiabatic approximation (11.38). Thus

$$
(\gamma_3 - 1)\left(\frac{\delta\rho}{\rho}\right)^* \delta\epsilon \approx (\gamma_3 - 1)[1 + \eta(\gamma_3 - 1)]\left|\frac{\delta\rho}{\rho}\right|^2 \geq 0 .
\tag{11.42}
$$

So the first term in the integrand of C always tends to *drive* oscillations. This is the so-called ϵ-mechanism. (However, it is not very important in most stars.) Physically, what happens is that when the star is compressed, the density and temperature increase, hence nuclear reactions increase and so more heat is generated.

The second term in the integrand of C is problematic in stars (such as the Sun) that possess convective regions. There, one needs to be able to estimate the perturbation to the convective flux of heat. In the absence of a reliable theory of convection, this is still a stumbling block. There has, however, been some success with stars where the flux is wholly radiative: the theory predicts the instability strip in the H-R diagram. Note that the κ-mechanism, discussed in the previous section, is contained within this term. Some further discussion of the various excitation mechanisms may be found in Unno *et al.* (1989).

Chapter 12

Nonradial Oscillations and Helioseismology

In the previous chapter we considered stellar oscillations where the motion was wholly in the radial direction. In this chapter we shall consider more general motions. The equilibrium structure about which the oscillations take place is still presumed to be spherically symmetric, but the velocity will now have a horizontal as well as a radial component, and the velocity and other perturbations will depend not only on r but also on θ and ϕ, where (r, θ, ϕ) are spherical polar coordinates.

One reason for studying nonradial oscillations of stars is that such modes of oscillation are observed (at very small amplitudes, of order $0.1\,\mathrm{m\,s^{-1}}$), on the Sun and are being used to make helioseismic studies of the interior of the Sun and hence to test our theories of stellar structure, the solar dynamo, *etc.* We shall address the subject of helioseismology explicitly later in the chapter. Other classes of pulsating stars are also known to exhibit nonradial oscillations (e.g. δ Scuti stars, rapidly oscillating Ap stars, slowly pulsating B stars, γ Doradus stars, and some white dwarf stars), as has already been illustrated in Fig. 11.1. Other solar-type stars are expected to have nonradial oscillations like the Sun's. Unlike the oscillations of classical pulsators, which are overstable, the oscillations in the Sun and solar-type stars are believed to be excited by turbulent generation of noise in their convection zones. There is now good evidence that some other solar-type stars pulsate, notably in the Sun's near-twin, α Centauri, as reported by Bouchy & Carrier (2001), making asteroseismology of these stars possible.

12.1 Nonradial Modes of Oscillation of a Star

It is helpful to define new notation to separate the horizontal component of the velocity from the vertical component and the angular dependence from the radial. The former separation we write in spherical polars as

$$\boldsymbol{u} = u_r \boldsymbol{e}_r + \boldsymbol{u}_h \qquad (\boldsymbol{e}_r \cdot \boldsymbol{u}_h = 0) \qquad (12.1)$$

for any $\boldsymbol{u}$. For the second separation, we define operators ∇_h, $\nabla_h \cdot$ etc. which contain all horizontal derivatives. In spherical polars (see Appendix B), we write

$$\nabla f \equiv \frac{\partial f}{\partial r} \boldsymbol{e}_r + \nabla_h f \,, \qquad \nabla \cdot \boldsymbol{u} \equiv \frac{1}{r^2} \frac{\partial}{\partial r} \left(r^2 u_r \right) + \nabla_h \cdot \boldsymbol{u}_h \qquad (12.2)$$

and, explicitly,

$$\nabla_h^2 f \equiv \frac{1}{r^2} \left[\frac{1}{\sin \theta} \frac{\partial}{\partial \theta} \left(\sin \theta \frac{\partial f}{\partial \theta} \right) + \frac{1}{\sin^2 \theta} \frac{\partial^2 f}{\partial \phi^2} \right] \,, \qquad (12.3)$$

for any f.

In this notation, Eqs. (2.32)–(2.34) become

$$\rho \frac{\partial u_r}{\partial t} = -\frac{\partial p'}{\partial r} - \rho' \frac{d\psi}{dr} - \rho \frac{\partial \psi'}{\partial r} \,, \qquad (12.4)$$

$$\rho \frac{\partial \boldsymbol{u}_h}{\partial t} = -\nabla_h p' - \rho \nabla_h \psi' \,, \qquad (12.5)$$

$$\frac{\partial \rho'}{\partial t} + \frac{1}{r^2} \frac{\partial}{\partial r} \left(r^2 \rho u_r \right) + \rho \nabla_h \cdot \boldsymbol{u}_h = 0 \,, \qquad (12.6)$$

$$\frac{1}{r^2} \frac{\partial}{\partial r} \left(r^2 \frac{\partial \psi'}{\partial r} \right) + \nabla_h^2 \psi' = 4\pi G \rho' \,, \qquad (12.7)$$

where Eq. (2.32) has been split into radial and horizontal components.

A final equation that we require is the energy equation, which for adiabatic oscillations is

$$\frac{\partial p'}{\partial t} + u_r \frac{dp}{dr} = c^2 \left(\frac{\partial \rho'}{\partial t} + u_r \frac{d\rho}{dr} \right) \,, \qquad (12.8)$$

where $c = (\gamma p/\rho)^{1/2}$ is, as usual, the adiabatic sound speed. By taking the horizontal divergence ($\nabla_h \cdot$) of Eq. (12.5), $\nabla_h \cdot \boldsymbol{u}_h$ can be eliminated between this and equation (12.6):

$$\frac{\partial}{\partial t} \left[\frac{\partial \rho'}{\partial t} + \frac{1}{r^2} \frac{\partial}{\partial r} \left(r^2 \rho u_r \right) \right] = \nabla_h^2 p' + \rho \nabla_h^2 \psi' \,. \qquad (12.9)$$

Thus we have four Eqs. (12.4), (12.7), (12.8), (12.9) in the four unknowns u_r, p', ρ' and ψ'.

Because the equilibrium state is independent of θ and ϕ, we may seek solutions that are separable in θ and ϕ: $u_r = R(r,t)S(\theta,\phi)$ and similarly for the other three variables. For this to work, we need the *same* angular dependence $S(\theta,\phi)$ for all four variables u_r, p', ρ' and ψ'. We also note that in the four governing equations, the only derivatives of θ and ϕ are in ∇_h^2. All terms must have the same angular dependence and so $\nabla_h^2 S$ must be proportional to S:

$$\nabla_h^2 S = -\frac{\Lambda}{r^2} S \, ,$$

where Λ is a constant. The $1/r^2$ factor is obvious from the expression (12.3) for ∇_h^2. Thus we must solve

$$\frac{1}{\sin\theta}\frac{\partial}{\partial\theta}\left(\sin\theta\frac{\partial S}{\partial\theta}\right) + \frac{1}{\sin^2\theta}\frac{\partial^2 S}{\partial\phi^2} = -\Lambda S \, . \tag{12.10}$$

This only has regular solutions if $\Lambda = l(l+1)$, where l is an integer (non-negative, without loss of generality). In that case, the regular solutions are the *spherical harmonics*

$$Y_l^m = P_l^{|m|}(\cos\theta)e^{im\phi} \qquad (m = -l, -(l-1), \cdots, l) \tag{12.11}$$

where $P_l^{|m|}(x)$ is an associated Legendre function and satisfies the associated Legendre equation

$$\frac{\mathrm{d}}{\mathrm{d}x}\left[(1-x^2)\frac{\mathrm{d}P_l^{|m|}}{\mathrm{d}x}\right] + \left(l(l+1) - \frac{m^2}{1-x^2}\right)P_l^{|m|}(x) = 0 \, . \tag{12.12}$$

Note that

$$\nabla_h^2 Y_l^m = -\frac{l(l+1)}{r^2}Y_l^m \, . \tag{12.13}$$

As for radial oscillations, we seek solutions with time dependence $\exp(-i\omega t)$. Then e.g.

$$u_r(r,\theta,\phi,t) = u_r(r)Y_l^m(\theta,\phi)e^{-i\omega t} \tag{12.14}$$

and similarly for the other three variables. Henceforth u_r, p', ρ', ψ' will be taken to denote just the radial dependence, as on the right-hand side of (12.14). The integers l and m are called the degree and the azimuthal order of the mode, respectively.

With the above separation of variables, Eqs. (12.4), (12.9), (12.7), (12.8) become, respectively,

$$-i\omega\rho u_r = -\frac{dp'}{dr} - \rho\frac{d\psi'}{dr} - \rho'g , \tag{12.15}$$

$$-i\omega\left[-i\omega\rho' + \frac{1}{r^2}\frac{d}{dr}\left(r^2\rho u_r\right)\right] = -\frac{l(l+1)}{r^2}\left(p' + \rho\psi'\right) , \tag{12.16}$$

$$\frac{1}{r^2}\frac{d}{dr}\left(r^2\frac{d\psi'}{dr}\right) - \frac{l(l+1)}{r^2}\psi' = 4\pi G\rho' , \tag{12.17}$$

$$\frac{1}{c^2}\left(-i\omega p' - \rho g u_r\right) = -i\omega\rho' + u_r\frac{d\rho}{dr} , \tag{12.18}$$

where g is the equilibrium gravitational acceleration: recall that, in the equilibrium state, $dp/dr = -\rho g$.

Using Eq. (12.18), ρ' can be eliminated from the other three equations. We thus arrive at

$$\frac{du_r}{dr} = -\left(\frac{2}{r} - \frac{g}{c^2}\right)u_r - \frac{-i\omega}{\rho c^2}\left(1 - \frac{S_l^2}{\omega^2}\right)p' - \frac{l(l+1)}{-i\omega r^2}\psi' , \tag{12.19}$$

$$\frac{dp'}{dr} = i\omega\rho\left(1 - \frac{N^2}{\omega^2}\right)u_r - \frac{g}{c^2}p' - \rho\frac{d\psi'}{dr} , \tag{12.20}$$

$$\frac{1}{r^2}\frac{d}{dr}\left(r^2\frac{d\psi'}{dr}\right) = 4\pi G\left(\frac{p'}{c^2} + \frac{\rho N^2}{-i\omega g}u_r\right) + \frac{l(l+1)}{r^2}\psi' , \tag{12.21}$$

where we have defined the Lamb frequency S_l by

$$S_l^2 \equiv \frac{l(l+1)c^2}{r^2} \tag{12.22}$$

and N is the Brunt-Väisälä frequency, given by Eq. (4.7). Both of S_l and N are functions of position within the star; they play an important role in determining the frequencies of nonradial oscillation of the star.

We note in passing that the horizontal component of the velocity can be determined from p' and ψ', using Eq. (12.5) and the definition (12.2) of ∇_h:

$$\boldsymbol{u}_h = u_h\left(\frac{\partial}{\partial\theta}\boldsymbol{e}_\theta + \frac{1}{\sin\theta}\frac{\partial}{\partial\phi}\boldsymbol{e}_\phi\right) , \tag{12.23}$$

where

$$u_h = \frac{1}{i\omega\rho r}\left(p' + \rho\psi'\right) . \tag{12.24}$$

Equations (12.19)–(12.21) form a fourth-order system in four unknowns (u_r, p', ψ' and $d\psi'/dr$). In addition, there is the unknown eigenvalue ω. To solve such a system requires four boundary conditions, plus a fifth which is just a normalization – for example, $u_r = 1$ at $r = R$. These boundary conditions are discussed in detail in e.g. Unno *et al.* (1989) and Christensen-Dalsgaard & Berthomieu (1991).

12.2 Mode Classification

For each value of l, one can try to solve Eqs. (12.19)–(12.21) plus the boundary conditions. It is only possible to solve the system of equations for certain discrete values of ω: these are the eigenvalues. In a Sturm–Liouville problem such as

$$y'' + \lambda^2 y = 0 \qquad y(0) = y(1) = 0 ,$$

(which has eigenvalues $\lambda_n = n\pi$ and eigenfunctions $y_n = \sin\lambda_n x$), the eigenfunctions have an increasing number of zeros as the eigenvalue increases. In the case of nonradial oscillations, one finds that the eigenvalues and eigenfunctions form a sequence such that, as ω gets large, the eigenfunctions have an increasing number of zeros, but also as $\omega \to 0$ the eigenfunctions gain more zeros also. One can label the modes with an integer n ($n = \ldots, -3, -2, -1, 0, 1, 2, 3, \ldots$) such that ω is an increasing function of n (at fixed l), that as $|n|$ increases the number of zeros in the eigenfunctions increases (except perhaps for small n), and that for large n, $|n|$ is the number of zeros in, for example, u_r. Thus any mode is described by three so-called quantum numbers, n, l and m: n is called the radial order of the mode.

Modes with $n > 0$ are called p modes: for these modes, the dominant restoring force is pressure. At high frequency these have the character of acoustic (sound) waves — see Section 2.5. Modes with $n < 0$ are called g modes: for these modes, the dominant resoring force is gravity. At low frequency these have the character of internal gravity waves. The $n = 0$ mode is called the f mode (f for fundamental). For large l, the f mode has the character of a surface gravity wave (see Section 2.6).

Notice that Eq. (12.19)–(12.21) depend on l, but the azimuthal order m does not appear. Hence, for given l, the eigenvalues and radial dependence of the eigenfunctions are independent of m. Thus one can label the eigenvalues with n and l only: $\omega = \omega_{nl}$. Figure 12.1 shows frequency ω as a function of l for a solar model. Frequencies of modes with the same value of n have been joined by continuous curves. Figure 12.2 shows eigenfunction u_r for a few modes of the same solar model.

12.3   The Cowling Approximation

It is not too difficult to solve the fourth-order system of differential equations numerically, but it is hard to gain much physical insight by so doing. Fortunately, ψ' can often be neglected, which reduces the system to second order. The circumstances in which ψ' can be neglected are when n or l is large. The eigenfunctions, and in particular ρ' then have many small-scale oscillations — either in radius or in the horizontal direction. The gravitational potential at a given point is an integral over the density distribution in space. The many positive and negative regions of ρ' tend to cancel out in ψ', leading to a very small perturbation to the gravitational potential. Mathematically, this can be seen from the solution (2.45) to the perturbed Poisson's equation (12.7). The solution with angular dependence Y_l^m that has the correct regular behaviour as $r \to 0$ and $r \to \infty$ is

$$\psi'(r) = \frac{4\pi G}{2l+1} \left[\frac{1}{r^{l+1}} \int_0^r \rho'(\tilde{r}) \tilde{r}^{l+2} \, d\tilde{r} + r^l \int_r^R \frac{\rho'(\tilde{r})}{\tilde{r}^{l-1}} \, d\tilde{r} \right] . \qquad (12.25)$$

Now, if l is large, $(\tilde{r}/r)^{l+2}$ is small in the first integral, and $(r/\tilde{r})^{l-1}$ is small in the second. Hence ψ' is small, compared with ρ'.

If n is large, then ρ' is a rapidly oscillating function of radius. The positive and negative contributions tend to cancel, so once again ψ' is small.

In the *Cowling approximation* (which is valid for large n or l) ψ' is neglected. Thus the equations become

$$\frac{du_r}{dr} = -\left(\frac{2}{r} - \frac{1}{\gamma H_p} \right) u_r - \frac{-i\omega}{\rho c^2} \left(1 - \frac{S_l^2}{\omega^2} \right) p' , \qquad (12.26)$$

$$\frac{dp'}{dr} = i\omega\rho \left(1 - \frac{N^2}{\omega^2} \right) u_r - \frac{1}{\gamma H_p} p' , \qquad (12.27)$$

where H_p, defined in Eq. (2.70), is the pressure scale height.

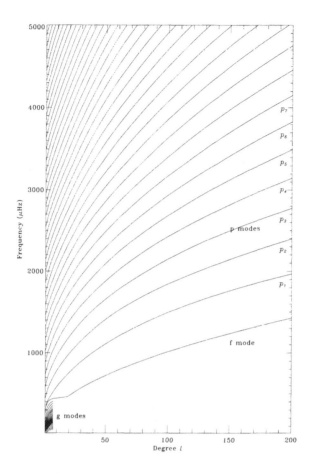

Fig. 12.1   The frequencies of modes of a solar model, as a function of degree l. All the mode frequencies fall along ridges, each ridge corresponding to a particular value of the radial order n. The p-mode ridges are labelled p_n where n is the radial order. Beneath the p modes is found the f-mode ridge. Beneath the f mode in frequency are the g modes: these are only shown in the figure for degrees less than 10, though they can exist for higher degrees also.

12.4   A Simplified Discussion of Nonradial Oscillations

We can gain insight into the nature of the solution of the nonradial wave equations by considering the relatively simple situation in which the derivatives of equilibrium quantities are neglected — we can expect this to be a reasonable approximation if the scale heights of the equilibrium quantities

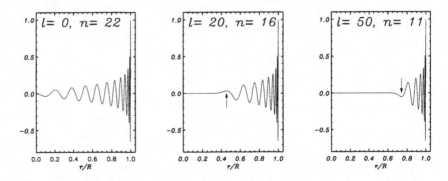

Fig. 12.2 Radial eigenfunctions of displacement for three p modes, with degree l and radial order n as indicated, as a function of fractional radius, for a solar model. All three modes have similar frequency around 3 mHz, i.e. periods of about 5 minutes. The eigenfunctions are very similar in the outer part of the Sun, because their frequencies are very similar. They penetrate to different depths, however: the penetration depth decreases with increasing degree. The location of the lower turning point at $r = r_t$, defined by Eq. (12.30), is indicated in each case by an arrow.

such as pressure are much greater than the scale on which the oscillations vary. We work in the Cowling approximation and so neglect the Eulerian perturbation to the gravitational potential. Then Eqs. (12.26), (12.27) become

$$\frac{\mathrm{d}u_r}{\mathrm{d}r} \simeq \frac{-i\omega}{\rho c^2}\left(\frac{S_l^2}{\omega^2} - 1\right)p' , \qquad \frac{\mathrm{d}p'}{\mathrm{d}r} \simeq i\omega\rho\left(1 - \frac{N^2}{\omega^2}\right)u_r . \qquad (12.28)$$

These can be combined into a single equation to give

$$\frac{\mathrm{d}^2 u_r}{\mathrm{d}r^2} \simeq \frac{\omega^2}{c^2}\left(\frac{S_l^2}{\omega^2} - 1\right)\left(1 - \frac{N^2}{\omega^2}\right)u_r \equiv -k_r^2 u_r , \qquad (12.29)$$

with solutions $\sin k_r r$, $\cos k_r r$. Clearly, the condition for these to be oscillatory (i.e. wavelike) is that $k_r^2 > 0$, i.e. either both $\omega^2 > S_l^2$ and $\omega^2 > N^2$ or both $\omega^2 < S_l^2$ and $\omega^2 < N^2$. Otherwise, the solution will be exponential. It is clear from this discussion, and indeed from Eq. (12.29), that k_r is the vertical wavenumber. Note that k_r is not a constant, but it varies only slowly with position.

Figure 12.3 shows the Lamb frequency S_l (at a few values of l) and buoyancy frequency N, for a solar model. In the solar case, the region where waves (so-called p modes) can propagate with $\omega^2 > S_l^2$, N^2 is between the surface and the point where $\omega^2 = S_l^2$. (Except for very low frequency and low degree, the buoyancy frequency is largely irrelevant for the Sun's p modes.) At this point in the star (at a radius $r = r_t$ say), it follows from Eq. (12.22) that

$$\frac{c(r_t)^2}{r_t^2} = \frac{\omega^2}{l(l+1)} . \tag{12.30}$$

If, as is indeed the case for observed solar p modes, $\omega^2 \gg N^2$, then k_r^2 can be approximated from Eq. (12.29) as

$$k_r^2 \simeq \frac{1}{c^2} \left(\omega^2 - S_l^2 \right) . \tag{12.31}$$

Let us compare this with the dispersion relation (2.51) for sound waves. Since, by Eq. (12.13), $\nabla_h^2 u_r = -l(l+1)u_r/r^2$, and recalling that for a plane wave $\nabla_h^2 \equiv -k_h^2$ (where k_h is the horizontal component of the wavenumber), we make the identification

$$\frac{l(l+1)}{r^2} \equiv k_h^2 . \tag{12.32}$$

Then Eq. (12.31) becomes $k_r^2 = \omega^2/c^2 - k_h^2$ or, re-arranging,

$$\omega^2 = c^2(k_r^2 + k_h^2) \equiv c^2|\boldsymbol{k}|^2$$

which is precisely the dispersion relation (2.51) for sound waves. Indeed, as was already anticipated, the high-frequency p modes have the character of acoustic waves.

The other regime where waves can propagate is where $\omega^2 < S_l^2, N^2$. For very low frequencies, noting that S_l^2 is generally bigger than N^2, we approximate k_r^2 from (12.29) as

$$k_r^2 \simeq \frac{\omega^2}{c^2} \frac{S_l^2}{\omega^2} \left(\frac{N^2}{\omega^2} - 1 \right) \simeq \frac{l(l+1)}{r^2} \left(\frac{N^2}{\omega^2} - 1 \right) . \tag{12.33}$$

Making the same identification for $l(l+1)/r^2$ as before, we thus have

$$\omega^2 = \frac{N^2 k_h^2}{k_r^2 + k_h^2} . \tag{12.34}$$

This is the dispersion relation for internal gravity waves. We shall not go further into that here, except to make two remarks. First, internal gravity waves are transverse waves, so the displacement of the fluid (and hence its velocity) is perpendicular to the direction of the wavenumber vector. By contrast, pure sound waves are longitudinal, i.e. the displacement and wavenumber are parallel. Secondly, if one considers the usual argument for establishing the Schwarzschild criterion for convective instability

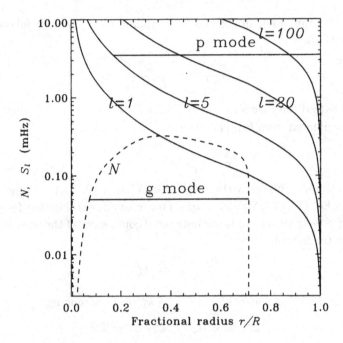

Fig. 12.3   Characteristic frequencies of a solar model: buoyancy frequency N (dashed curve) and Lamb frequency S_l for various values of mode degree l (solid curves). The two horizontal lines show the approximate cavity in which a 0.5 mHz g mode and an $l = 20$, 3.5 mHz p mode would propagate.

(Section 4.1.1), one finds that a blob of fluid displaced vertically in a convectively stable medium will oscillate with a frequency equal to N: this is precisely what (12.34) gives for the case $k_r = 0$.

12.5   A More General Asymptotic Expression

In Section 12.4 it was assumed that spatial derivatives of equilibrium quantities could be neglected entirely. We now relax that assumption, to obtain a more accurate description of the waves. We shall still work within the Cowling approximation.

It turns out to be mathematically convenient to work with $\nabla \cdot \boldsymbol{u}$ rather than u_r or p'. It follows from the adiabatic energy equation (12.8) and

Eq. (2.33) that, since u_r and p' are proportional to $Y_l^m \exp(-i\omega t)$,

$$\nabla \cdot \boldsymbol{u} = \chi(r) Y_l^m(\theta, \phi) e^{-i\omega t} \tag{12.35}$$

where

$$\chi(r) = \frac{-1}{\rho c^2} \left(-i\omega p' + u_r \frac{dp}{dr} \right) . \tag{12.36}$$

In the following we shall neglect terms arising from the spherical geometry (e.g. the $2/r$ term in Eq. (12.26)) and also the spatial variation of gravity. These are adequate approximations for the Sun's convective envelope and the outer part of the radiative interior, for example. Then, with the usual identification $l(l+1)/r^2 \equiv k_h^2$, Eqs. (12.26), (12.27) become

$$\frac{du_r}{dr} = \chi + \frac{-i\omega}{\rho \omega^2} k_h^2 p' , \tag{12.37}$$

$$\frac{dp'}{dr} = i\omega\rho \left(1 + \frac{g}{\omega^2} \frac{d\ln\rho}{dr} \right) u_r + i\omega \frac{\rho g}{\omega^2} \chi . \tag{12.38}$$

With some algebra, one can eliminate u_r and p' between Eqs. (12.36), (12.37) and (12.38) to arrive finally at

$$\frac{d^2\chi}{dr^2} + \left(\frac{2}{c^2} \frac{dc^2}{dr} + \frac{1}{\rho} \frac{d\rho}{dr} \right) \frac{d\chi}{dr} \tag{12.39}$$
$$+ \left[\frac{1}{\gamma} \frac{d^2\gamma}{dr^2} - \frac{2}{\gamma} \frac{d\gamma}{dr} \frac{\rho g}{p} + k_h^2 \left(\frac{N^2}{\omega^2} - 1 \right) - \frac{1}{\rho} \frac{d\rho}{dr} \frac{1}{\gamma} \frac{d\gamma}{dr} + \frac{\rho\omega^2}{\gamma p} \right] \chi = 0 .$$

To make further progress, we make a change of dependent variable to eliminate the first-derivative term and reduce the equation to standard form. The trick is to make the substitution $\chi = \Phi v$, where the function v is chosen such that no $d\Phi/dr$ term is left in the equation. Now

$$\chi' = \Phi' v + \Phi v' \qquad \chi'' = \Phi'' v + 2\Phi' v' + \Phi v'' .$$

Substituting these into Eq. (12.39), the coefficient of Φ' that we wish to make zero is

$$2v' + \left(\frac{2}{c^2} \frac{dc^2}{dr} + \frac{1}{\rho} \frac{d\rho}{dr} \right) v = 0 .$$

This can easily be integrated to find (up to an arbitary multiplicative constant) that

$$v = \rho^{-1/2} c^{-2} .$$

With this change of variable, Eq. (12.39) becomes

$$\frac{d^2\Phi}{dr^2} + \left[k_h^2\left(\frac{N^2}{\omega^2} - 1\right) + \frac{\omega^2}{c^2} - \frac{1}{2}\frac{d}{dr}(H_\rho^{-1}) - \frac{1}{4}H_\rho^{-2}\right]\Phi = 0, \quad (12.40)$$

where H_ρ is the density scale height: $H_\rho^{-1} = -\rho^{-1}d\rho/dr$. Let

$$\omega_c^2 = \frac{c^2}{4H_\rho^2}\left(1 - 2\frac{dH_\rho}{dr}\right); \quad (12.41)$$

$\omega_c(r)$ is called the acoustic cut-off frequency. Then Eq. (12.40) can be written in standard form as

$$\frac{d^2\Phi}{dr^2} + k_r^2(r)\Phi = 0, \quad (12.42)$$

where

$$k_r^2(r) = \frac{\omega^2 - \omega_c^2}{c^2} - \frac{S_l^2}{c^2}\left(1 - \frac{N^2}{\omega^2}\right). \quad (12.43)$$

(Deubner & Gough 1984: see also Gough 1993).

From our previous simplified discussion, we recognise k_r as a local radial wavenumber. If k_r^2 is positive, we have an oscillatory solution, corresponding to propagating waves, whereas if k_r^2 is negative the solution is exponential. Now the acoustic cut-off frequency depends on the density scale height. Near the surface of the Sun, for example, H_ρ gets small and so H_ρ^{-1} gets large. Assuming that other terms in k_r^2 do not matter near the surface, we see that the mode cannot propagate in a region where $\omega < \omega_c$. This causes waves propagating through the solar interior to be reflected near the surface of the Sun.

Expression (12.43) can be factorized as

$$k_r^2 = \frac{\omega^2}{c^2}\left(1 - \frac{\omega_+^2}{\omega^2}\right)\left(1 - \frac{\omega_-^2}{\omega^2}\right), \quad (12.44)$$

where ω_+^2 and ω_-^2 are roots of the quadratic

$$\left(\omega^2\right)^2 - \left(S_l^2 - \omega_c^2\right)\omega^2 + N^2S_l^2 = 0$$

i.e.

$$\omega_\pm^2 = \frac{1}{2}\left\{S_l^2 - \omega_c^2 \pm \left[\left(S_l^2 - \omega_c^2\right)^2 - 4N^2S_l^2\right]^{1/2}\right\} \quad (12.45)$$

(Deubner & Gough 1984). In the deep solar interior, ω_c^2 is very small. Since N^2 is generally small compared with S_l^2 (see Fig. 12.3), the roots (12.45) are roughly $\omega_+^2 = S_l^2$ and $\omega_-^2 = N^2$. For locally propagating (oscillatory) solutions, k_r^2 must be positive; and from (12.44) we thus see that this means that $\omega^2 > S_l^2$, N^2 or $\omega^2 < S_l^2$, N^2: cf. the discussion following Eq. (12.29).

If k_r^2 in Eq. (12.42) were constant, the solution Φ would be $\sin k_r r$ or $\cos k_r r$ (or equivalently $\exp(\pm i k_r r)$). In fact k_r^2 is not constant: it is a function of r. However, we can get a similar approximate solution if k_r^2 varies only *slowly* with position. By slowly, we mean that the length scale over which k_r varies is much greater than the radial wavelength $2\pi/k_r$ of the wave-like solution. Then, using JWKB theory (see Appendix D), we can write down the approximate solution of (12.42) as

$$\Phi \propto \left[k_r^2(r)\right]^{-1/4} \exp\left[\pm i \int^r k_r(r')\,\mathrm{d}r'\right] . \tag{12.46}$$

Note that this would reduce to a purely sinusoidal solution if k_r^2 were constant.

We can expect (12.46) to be a good approximation for high-order ($n \gg 1$) p modes. For these modes $\omega^2 \gg N^2$, and so by Eq. (12.44)

$$k_r^2 \simeq \frac{\omega^2}{c^2} - \frac{S_l^2}{c^2} = \frac{\omega^2}{c^2} - \frac{L^2}{r^2} \tag{12.47}$$

where $L^2 = l(l+1)$. Thus, finally, we obtain an approximate solution for the radial dependence χ of $\nabla \cdot \boldsymbol{u}$, Eq. (12.35):

$$\chi = \frac{A}{\rho^{1/2}c^2}\left(\frac{\omega^2}{c^2} - \frac{L^2}{r^2}\right)^{-1/4} \cos\left[\omega \int_{r_t}^r \left(1 - \frac{L^2 c^2}{\omega^2 r^2}\right)^{1/2} \frac{\mathrm{d}r}{c} + \phi\right] \tag{12.48}$$

where A and ϕ are constants, and we have chosen the lower limit of integration to be the radius of the lower turning point of the mode, where the integrand vanishes.

12.6 Helioseismology: The Duvall Law

We have established how the resonant frequencies of nonradial stellar oscillations depend upon conditions in the stellar interior. If the frequencies of many such modes can be measured accurately, then it becomes possible to perform stellar seismology: to make inferences about conditions in the interior from the measured frequencies. Because the Sun is close, we are able to resolve its disk very well and have been able to detect many nonradial resonant oscillations. The practice of inferring conditions in the solar interior from its resonant oscillation frequencies is known as helioseismology. A good review of helioseismology may be found in Christensen-Dalsgaard (2002). The only modes to have been unambiguously identified on the Sun are p and f modes: g-mode identifications are still a matter of controversy. We shall consider g modes only in Section 12.8.

Applying JWKB theory (Appendix D) to Eq. (12.42), we have that

$$(n + \epsilon)\pi = \int_{r_t}^{R_t} k_r \, dr \,, \tag{12.49}$$

where r_t and R_t are respectively the radii of the lower and upper turning points of the mode (where k_r vanishes), n is an integer and ϵ is a constant. In words, apart from allowance for a phase ($\epsilon\pi$) at the turning points, an integral number of half-wavelengths fit between the r_t and R_t. Substituting from (12.43), and neglecting N^2/ω^2 (which is small for high-frequency p modes), we deduce that

$$\frac{(n + \epsilon)\pi}{\omega} = \int_{r_t}^{R_t} \left(1 - \frac{\omega_c^2}{\omega^2} - \frac{L^2 c^2}{\omega^2 r^2} \right)^{1/2} \frac{dr}{c} \,, \tag{12.50}$$

where $L = \sqrt{l(l+1)}$.

In the solar interior, $\omega_c^2 \ll \omega^2$ (the acoustic cut-off frequency being important only near the surface). On the other hand, near the surface $L^2 c^2/\omega^2 r^2 \ll 1$, so this term is relatively unimportant there. Moreover, the upper turning point is very close to the photosphere. Hence it can be shown that the upper limit of integration can be moved to $r = R$ and the ω_c contribution absorbed into a frequency-dependent phase function:

$$\frac{(n + \alpha(\omega))\pi}{\omega} = \int_{r_t}^{R} \left(1 - \frac{L^2 c^2}{\omega^2 r^2} \right)^{1/2} \frac{dr}{c} \equiv F(\omega/L) \,. \tag{12.51}$$

The only aspect of the equilibrium model that enters Eq. (12.51) is the sound speed (reasonably, since we have already identified the high-frequency p modes as acoustic waves). Using (12.51), the internal sound speed in the Sun can be inferred from the observed frequencies.

First, we note that the right-hand side of (12.51) is a function of the mode-dependent quantities ω and L only in the combination ω/L (note that r_t is a function of ω/L also). Suppose then that we observe many modes and measure their frequencies. Their values of L can be determined from observing their horizontal wavelength, and their n values, though not directly observable, can be inferred by comparing the observed frequencies with those of theoretical solar models. One can then plot $(n+\alpha)\pi/\omega$ against ω/L, choosing $\alpha(\omega)$ so that the points lie on a single curve. This curve is $F(\omega/L)$. Henceforth we write $\omega/L \equiv w$. Equation (12.51) is sometimes called the Duvall law, after Duvall (1982).

Having thus determined $F(w)$ observationally, one can obtain $c(r)$ by *inverting*

$$F(w) = \int_{r_t}^{R} \left(1 - \frac{L^2 c^2}{\omega^2 r^2}\right)^{1/2} \frac{dr}{c} . \tag{12.52}$$

If we define a new variable $a = c/r$ we can rewrite (12.52) as

$$F(w) = -\int_{a_s}^{w} \left(1 - \frac{a^2}{w^2}\right)^{1/2} a^{-1} \frac{d\ln r}{da} \frac{dr}{c} , \tag{12.53}$$

where a_s is the value of a at $r = R$. This is an integral of the Abel type and can be inverted analytically. The basic idea is that if

$$I(y) = \int_{0}^{y} (y^2 - x^2)^{-1/2} f(x) dx \tag{12.54}$$

then

$$\int_{0}^{z} (z^2 - y^2)^{-1/2} I(y) y \, dy = \int_{0}^{z} dy \int_{0}^{y} dx (z^2 - y^2)^{-1/2} (y^2 - x^2)^{-1/2} y f(x)$$

$$= \int_{0}^{z} dx \int_{x}^{z} dy (z^2 - y^2)^{-1/2} (y^2 - x^2)^{-1/2} y f(x) .$$

Figure 12.4 shows schematically how the necessary re-ordering of the integrations over x and y is achieved. Using the substitution $y^2 = z^2 \sin^2\theta + x^2 \cos^2\theta$, one finds that $y \, dy = (z^2 - x^2) \sin\theta \cos\theta \, d\theta$ and so

$$\int_{0}^{z} (z^2 - y^2)^{-1/2} I(y) y \, dy$$

$$= \int_{0}^{z} dx \int_{0}^{\pi/2} d\theta \frac{(z^2 - x^2) \sin\theta \cos\theta f(x)}{(z^2 - x^2)^{1/2} \cos\theta (z^2 - x^2)^{1/2} \sin\theta} = \frac{\pi}{2} \int_{0}^{z} f(x) dx .$$

Hence, differentiating,

$$f(z) = \frac{2}{\pi} \frac{d}{dz} \int_{0}^{z} (z^2 - y^2)^{-1/2} I(y) y \, dy . \tag{12.55}$$

Thus, if $I(y)$ is known for all y, one can determine $f(z)$ for all z.

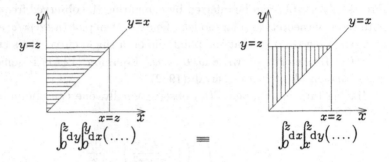

Fig. 12.4 Schematic illustration of the re-ordering of the two integrations in the inversion discussed in Section 12.6. In the initial ordering, on the left-hand side, the x-integral is performed first, along the horizontal strips, and then the y-integral is performed. When the order of the integrations is swapped over, the y-integrals (along the vertical strips) is performed first. The limits on the integrals consequently change, as indicated in the figure.

Based on this, with some work, one can obtain the sound speed in the Sun implicitly from (12.53) as

$$r = R \exp\left[\frac{-2}{\pi} \int_{a_s}^{a} \left(w^{-2} - a^{-2}\right)^{-1/2} \frac{dF}{dw} dw\right] . \tag{12.56}$$

Hence one can plot r as a function of $a \equiv c/r$ and thus, if one picks any value of r, one can determine $c(r)$ there. This procedure is discussed in Gough (1993).

12.7 Tassoul's Formula

We now derive an approximate formula for low-degree modes. Since modes penetrate down to their lower turning point, defined to be where $c(r_t)/r_t = \omega/L$, and given that c/r increases with decreasing r, it is clear that modes with small values of l penetrate most deeply. The very lowest degree modes can tell us about conditions in the energy-generating core, and are thus of great interest to helioseismologists.

Equation (12.51) can be rewritten as

$$\frac{(n + \alpha(\omega))\pi}{\omega} = \int_{0}^{R} \frac{dr}{c} - I \tag{12.57}$$

where

$$I = \int_0^R \frac{dr}{c} - \int_{r_t}^R \left(1 - \frac{L^2 c^2}{\omega^2 r^2}\right)^{1/2} \frac{dr}{c}$$

$$= \int_0^{r_t} \frac{dr}{c} + \int_{r_t}^R \left[1 - \left(1 - \frac{L^2 c^2}{\omega^2 r^2}\right)^{1/2}\right] \frac{dr}{c} \equiv I_1 + I_2 . \quad (12.58)$$

For observed low-degree p modes (frequencies $\nu \equiv \omega/2\pi \simeq 3\text{mHz}$), r_t is close to the centre. Thus $I_1 \simeq r_t/c(r_t) = L/\omega$. To approximate I_2, note that the integrand is small except when $r \simeq r_t$; hence we will replace $c(r)$ with $c(r_t)$ and also extend the upper limit of integration to infinity. With the change of variable $u = Lc(r_t)/(\omega r)$, one can show that

$$I_2 \simeq -\frac{L}{\omega} \int_1^0 \left[1 - (1 - u^2)^{1/2}\right] \frac{du}{u^2} = \frac{L}{\omega} \left(\frac{\pi}{2} - 1\right) . \quad (12.59)$$

Thus

$$I = I_1 + I_2 \simeq \frac{L}{\omega} \frac{\pi}{2} .$$

Substituting this back into Eq. (12.57) and re-arranging gives

$$\frac{(n + L/2 + \alpha)\pi}{\omega} = \int_0^R \frac{dr}{c} \equiv T$$

where T is the sound travel time from the centre to the surface of the Sun. Writing this in terms of the cyclic frequency $\nu = \omega/2\pi$, and defining $\nu_0 = (2T)^{-1}$, we have that

$$\nu = \left(n + \frac{1}{2}L + \alpha\right)\nu_0 . \quad (12.60)$$

This formula shows that the frequencies of low-degree modes with the same value of l but different values of n are roughly equally spaced, with a separation equal to the reciprocal of the total sound travel time from the surface to the centre and back. For more details, see Tassoul (1980), Gough (1993).

A more accurate formula can be obtained with rather more care in the approximations:

$$\nu \simeq \left(n + \frac{1}{2}\tilde{L} + \alpha\right) - \left(A\tilde{L}^2 - \delta\right)\frac{\nu_0^2}{\nu} + O\left(\nu_0^4/\nu^3\right) , \quad (12.61)$$

where $\tilde{L} = (l + 1/2)$, ν_0 is as above, α and δ are constants that depend (in particular) on the near-surface stratification, and constant A is given approximately by

$$A = \frac{1}{(2\pi)^2 \nu_0} \left(\frac{C(R)}{R} - \int_{r_t}^{R} \frac{1}{r}\frac{dc}{dr}dr \right) . \qquad (12.62)$$

Equation (12.61) is important not only for studies of the deep solar interior, but also for its potential to enable us to perform seismology on more distant stars. For in general we cannot expect to be able to detect high-degree oscillations on distant stars, but we may see low-degree oscillations, to which this formula is applicable. The dominant term is the first one, and we note that the frequencies of modes with the same value of l but adjacent values of n will be separated by an amount ν_0: for the Sun this is about $140\,\mu$Hz, and is sometimes called the large separation (in contrast to what comes next). We note secondly that the dominant first term will be equal for two modes with order and degree (n, l) and $(n - 1, l + 2)$. Thus there is a near-degeneracy between the $l = 0$ and $l = 2$ modes, say, and between $l = 1$ and $l = 3$ modes. The degeneracy is not precise, because of the next terms in the asymptotic expansion, which leads to a difference between $\nu_{n,l}$ and $\nu_{n-1,l+2}$ equal to $A(4l + 6)$. For the Sun this is about $9(2l/3 + 1)\,\mu$Hz. This is sometimes called the small separation.

The large separation is related to the sound travel time throughout the whole star. By contrast, the small separation is particularly sensitive to conditions in the core. These quantities change during the lifetime of a star, and are different for stars of different mass. Under the assumption that other aspects of the stellar physics are correctly modelled, the large and small separations can be used to estimate the masses and ages of stars like the Sun whose oscillation spectra exhibit low-degree p modes.

12.8 Asymptotics of g Modes

For g modes, $\omega^2 < N^2$ and $\omega^2 \ll S_l^2$. Thus the dispersion relation (12.43) can be approximated as

$$k_r^2 \simeq \frac{S_l^2}{c^2}\left(\frac{N^2}{\omega^2} - 1\right) = \frac{L^2}{r^2}\left(\frac{N^2}{\omega^2} - 1\right) .$$

Thus

$$(n + \epsilon)\pi \simeq \int_{r_1}^{r_2} \frac{L}{r}\left(\frac{N^2}{\omega^2} - 1\right)^{1/2} dr , \qquad (12.63)$$

where r_1 and r_2 are the radii at which k_r^2 vanishes. For low-frequency g modes, we thus have that the period $2\pi/\omega$ is given by

$$\frac{2\pi}{\omega} = \frac{2\pi^2(n+\epsilon)}{\displaystyle\int_{r_1}^{r_2} \frac{L}{r} N \mathrm{d}r}. \qquad (12.64)$$

This formula predicts a roughly regular spacing in *period* between g modes with the same l but different n. This regular pattern has been used to search for solar g modes. Such modes have not yet been unambiguously detected, though there have been claimed detections. That it is difficult to see g modes is not surprising, since N^2 is essentially zero in the Sun's convection zone and so the modes must have a long exponential tail to the observable surface. It is hoped that observations from space in the near future will have sufficient sensitivity to small velocities that g modes might be observed. Being trapped deep in the interior, they would be an invaluable tool for probing the energy-generating core.

12.9 Probing the Sun's Internal Rotation

Thus far, we have assumed that the nonradial oscillations take place in a spherically symmetric, nonrotating star. In that case, the frequencies are independent of the azimuthal order m. Rotation raises this degeneracy. By measuring the differences in frequency between modes with the same values of n and l but with different values of m, we can probe the internal solar rotation (and possibly other asymmetries, though we shall not consider that here).

The Sun is a slow rotator: its rotation period is of the order of one month (compared with the typical p-mode periods of 5 minutes) and the fractional difference between its polar and equatorial radii is only of order 10^{-5}. Thus we will use a perturbation approximation to estimate the effect of rotation on the p-mode frequencies, and we shall neglect terms quadratic (or higher) in the angular velocity Ω.

Consider small perturbations about an equilibrium state in which the star is spherically symmetric and rotating with a steady velocity $\Omega \times r$. (Centrifugal forces would distort the star from spherical symmetry, but this effect would be quadratic in the rotation rate, and we shall neglect second-order effects of rotation throughout this analysis.) We take the polar axis of our coordinate system such that the angular velocity is in that direction, with magnitude Ω. We shall consider the total velocity of the fluid to be $\boldsymbol{\Omega} \times \boldsymbol{r} + \boldsymbol{u}$, where $\boldsymbol{u}$ is a small perturbation; and all other notation is as before. In particular we seek perturbations proportional to $\exp(-i\omega t)$, and we shall work in the Cowling approximation. Then it can be shown (Lynden-Bell & Ostriker 1967) that

$$\mathcal{L}\boldsymbol{u} + \rho\omega^2\boldsymbol{u} = -2i\omega\rho\Omega\frac{\partial\boldsymbol{u}}{\partial\phi}, \qquad (12.65)$$

where

$$\mathcal{L}\boldsymbol{\xi} \equiv \nabla\left[\gamma p\nabla \cdot \boldsymbol{u} + \boldsymbol{u} \cdot \nabla p\right] - \frac{1}{\rho}\nabla \cdot (\rho\boldsymbol{u})\nabla p : \qquad (12.66)$$

cf. Eq. (11.1). With appropriate boundary conditions, $\mathcal{L}$ is a self-adjoint operator.

We shall denote the eigenfunction corresponding to quantum numbers (n, l, m) as $\boldsymbol{u}_{nlm}$. In the absence of rotation, the star is spherically symmetric and modes with quantum numbers (n, l, m) and (n, l, m') have identical frequencies. Thus the frequencies ω_{nlm} are degenerate, with degeneracy $2l + 1$. Rotation raises this degeneracy. Given that the rotation is slow, we can estimate the frequencies in the rotating star using first-order degenerate perturbation theory. The details of degenerate perturbation theory can be found in many introductory books on quantum mechanics, e.g. Schiff (1968). We write the frequency as $\omega = \omega_{nl} + \omega_1$, and the eigenfunction as

$$\boldsymbol{u} = \sum_{m'} c_{m'}\boldsymbol{u}_{nlm'} \qquad (12.67)$$

where the coefficients $c_{m'}$ are as yet unknown. This is an essential point about degenerate perturbation theory: the zero-order eigenfunction is a linear combination of the degenerate unperturbed eigenfunctions, the exact form of which is only determined by considering the first-order perturbation. We shall assume that the perturbation ω_1 to the frequency is of the same order as Ω, and we shall work only to first order in Ω.

Of course, the unperturbed eigenfunction $\boldsymbol{u}_{nlm}$ satisfies

$$\mathcal{L}\boldsymbol{u}_{nlm} + \rho\omega_{nl}^2\boldsymbol{u}_{nlm} = 0. \qquad (12.68)$$

We shall denote the integral $\int \ldots dV$ over the interior of the star by $\langle \ldots \rangle$.

Now, substituting expression (12.67) into Eq. (12.65), multiplying by

u_{nlm}^* and integrating over the interior of the star yields

$$\sum_{m'} \langle u_{nlm}^* \cdot \mathcal{L} u_{nlm'} \rangle + (\omega_{nl}^2 + 2\omega_{nl}\omega_1)\langle \rho u_{nlm}^* \cdot u_{nlm'} \rangle$$

$$= -2i\omega_{nl}\langle \Omega u_{nlm}^* \cdot \frac{\partial}{\partial \phi} u_{nlm'} \rangle . \qquad (12.69)$$

By self-adjointness,

$$\langle u_{nlm}^* \cdot \mathcal{L} u_{nlm'} \rangle = \langle u_{nlm'} \cdot (\mathcal{L} u_{nlm})^* \rangle = -\omega_{nl}^2 \langle \rho u_{nlm'}^* \cdot u_{nlm} \rangle$$

using Eq. (12.68). Also we note that, assuming Ω to be independent of ϕ, the angular dependence of the integrand on the right of Eq. (12.69) is $\exp(i(m'-m)\phi)$. Hence, when the integral over ϕ from 0 to 2π is performed, only the term with $m' = m$ will survive. The same is true of the integral $\langle \rho u_{nlm}^* \cdot u_{nlm'} \rangle$. Thus one finds that

$$\sum_{m'} (A_{mm'} - \omega_1 \delta_{mm'}) c_{m'} = 0 , \qquad (12.70)$$

where A is a diagonal matrix with diagonal elements

$$A_{mm} = \frac{-2i\omega_{nl}\langle \rho \Omega u_{nlm}^* \cdot \partial u_{nlm}/\partial \phi \rangle}{\langle \rho u_{nlm}^* \cdot u_{nlm} \rangle} .$$

Equation (12.70) is a matrix eigenvalue equation which determines the eigenvalues ω_1 and the eigenvectors (the elements of which are the coefficients c_m. It is of a particularly simple form, since the matrix is diagonal: the eigenvalues are just equal to the elements A_{mm}, and the eigenvectors are simply given by one of the c_m being unity and the others zero. There are $2l + 1$ such eigensolutions.

Thus each eigenfunction of the rotating star corresponds to a particular (n, l, m), and the corresponding frequency is

$$\omega_{nlm} = \omega_{nl} + A_{mm} .$$

Now, using Eq. (12.23), u_{nlm} can be written in the form

$$u_{nlm} = u_r(r)Y_l^m e_r + u_h(r)\left(\frac{\partial Y_l^m}{\partial \theta}e_\theta + \frac{1}{\sin\theta}\frac{\partial Y_l^m}{\partial \phi}e_\phi\right) .$$

Assuming that the spherically symmetric structure of the Sun (or other star) is accurately known, the eigenfunctions u_{nlm} can be calculated and so one can calculate *kernels* $K_{nlm}(r, \theta)$ such that

$$A_{mm} = \int_0^R \int_0^\pi K_{nlm}\Omega \, dr \, d\theta \, .$$

Then

$$\omega_{nlm} - \omega_{nl0} \equiv \omega_{nlm} - \omega_{nl} = \int_0^R \int_0^\pi K_{nlm}(r, \theta)\Omega(r, \theta) \, dr \, d\theta \, . \quad (12.71)$$

The left-hand side of Eq. (12.71), which is called the rotational splitting, is measurable for many modes of oscillation of the Sun. The kernels on the right-hand side, which show how the left-hand side is a weighted average of the underlying rotation rate, are also known. Provided there is sufficient variety among the kernels in the way they sample the interior, it is then possible to make inferences about how the rotation rate varies inside the Sun. The results on solar internal rotation shown in Fig. 3.1 were obtained in this way.

There are various so-called inversion methods for taking sets of data constraints of the form of (12.71) with known kernel functions K_{nlm} and making inferences about the unknown function Ω. Several are described in Gough (1985) and Christensen-Dalsgaard (2002). Such methods can also be used to make inferences about the internal structure of the Sun, by linearizing structural differences and frequency differences with respect to a known solar model: such methods complement the asymptotic techniques described in Section 12.6 and are the basis for many of the results from helioseismology.

Finally, we note that if the rotation rate Ω is a function only of radius r, then it follows (after some algebra) that

$$\omega_{nlm} - \omega_{nl0} = m \int_0^R K_{nl}(r)\Omega(r) \, dr \, , \quad (12.72)$$

where

$$K_{nl}(r) = \frac{\left[(u_r - u_h)^2 + (l(l+1) - 2) u_h^2\right] \rho r^2}{\int_0^R u_r^2 + l(l+1)u_h^2 \, dr} \, . \quad (12.73)$$

Note that in this case the frequency splitting is simply proportional to m. Deviations from such a linear relationship in the observed splittings are therefore indicative of latitudinal variation in the Sun's internal rotation.

In the particular case of rigid-body rotation (Ω constant), we have from Eqs. (12.72) and (12.73) that

$$\omega_{nlm} - \omega_{nl0} = m(1 - C_{nl})\Omega .$$

For high-frequency p modes, u_h^2 is everywhere much smaller than $u_r^2 + l(l+1)u_h^2$. It then follows by inspecting Eqs. (12.72) and (12.73) that C_{nl}, which is known as the Ledoux constant, is small. It represents the small contribution from Coriolis forces to the overall frequency splitting due to rotation.

Appendix A

Useful Constants and Quantities

A.1 Fundamental Physical Constants

a	radiation constant	7.57×10^{-16} J m^{-3} K^{-4}
c	speed of light	3.00×10^{8} m s^{-1}
e	magnitude of charge on electron	1.60×10^{-19} C
G	gravitational constant	6.67×10^{-11} N m^2 kg^{-2}
h	Planck's constant	6.63×10^{-34} J s
k	Boltzmann's constant	1.38×10^{-23} J K^{-1}
m_e	mass of electron	9.11×10^{-31} kg
m_H	mass of hydrogen atom	1.67×10^{-27} kg
N_A	Avogadro's number	6.02×10^{23} mol^{-1}
σ	Stefan–Boltzmann constant	5.67×10^{-8} W m^{-2} K^{-4}
ϵ_0	permittivity of free space	8.85×10^{-12} farad m^{-1}
μ_0	permeability of free space	$4\pi \times 10^{-7}$ henry m^{-1}
$\mathcal{R}$	gas constant (k/m_H)	8.26×10^{3} J K^{-1} kg^{-1}

A.2 Astronomical Quantities

$L_\odot$	solar luminosity	3.83×10^{26} W
$M_\odot$	solar mass	1.99×10^{30} kg
$R_\odot$	solar radius	6.96×10^{8} m
$T_{\mathrm{eff}\odot}$	Sun's effective temperature	5770 K
AU	astronomical unit	1.50×10^{11} m
pc	parsec	3.09×10^{16} m

A.3   Cartesian Tensors: Index Notation and Summation Convention

We denote the ith component of vector $\boldsymbol{u}$ (say) by u_i. We denote the ith component of the position vector by x_i. Scalars have no indices, vectors have a single index, second-rank tensors have two indices, third-rank tensors have three indices, etc. The indices $(i, j, \dots)$ can take possible values 1, 2 and 3 in three-dimensional space.

Two useful quantities are

$$\delta_{ij} = \begin{cases} 1 & \text{if } i = j \\ 0 & \text{if } i \neq j \end{cases} \tag{A.1}$$

and

$$\epsilon_{ijk} = \begin{cases} 1 & \text{if } (i,j,k) \text{ is a cyclic permutation of } (1,2,3) \\ -1 & \text{if } (i,j,k) \text{ is an anticyclic perm. of } (1,2,3) \\ 0 & \text{otherwise} \end{cases} \tag{A.2}$$

Note that the indices of ϵ_{ijk} can be cyclically permuted without changing its value.

If the same index is repeated, then it means that it must be summed over values 1, 2 and 3. Thus $u_i v_i \equiv u_1 v_1 + u_2 v_2 + u_3 v_3$ is the scalar product of $\boldsymbol{u}$ and $\boldsymbol{v}$, and $\epsilon_{ijk}(\partial u_k/\partial x_j)$ is the ith component of the vector $\nabla \times \boldsymbol{u}$. It is useful to note that $\delta_{ij} u_j = \delta_{i1} u_1 + \delta_{i2} u_2 + \delta_{i3} u_3 = u_i$.

A very useful identity, which facilitates the proof of various standard identities for vectors and vector calculus, is

$$\epsilon_{ijk}\epsilon_{klm} = \delta_{il}\delta_{jm} - \delta_{im}\delta_{jl} . \tag{A.3}$$

For example, with "ith component of" understood,

$$\begin{aligned}
\nabla \times (\boldsymbol{u} \times \boldsymbol{v}) &= \epsilon_{ijk} \frac{\partial}{\partial x_j}(\epsilon_{klm} u_l v_m) \\
&= \frac{\partial}{\partial x_j}[(\delta_{il}\delta_{jm} - \delta_{im}\delta_{jl}) u_l v_m] \\
&= \frac{\partial}{\partial x_j}(u_i v_j) - \frac{\partial}{\partial x_j}(u_j v_i) \\
&= \boldsymbol{u}(\nabla \cdot \boldsymbol{v}) + \boldsymbol{v} \cdot \nabla \boldsymbol{u} - \boldsymbol{v}(\nabla \cdot \boldsymbol{u}) - \boldsymbol{u} \cdot \nabla \boldsymbol{v} . \tag{A.4}
\end{aligned}$$

Vector Calculus in Spherical and Cylindrical Polar Coordinates

B.1 Cylindrical Polar Coordinates (ϖ, ϕ, z)

In cylindrical polar coordinates (ϖ, ϕ, z), where coordinate ϖ is the radial distance from the axis, let ψ be a scalar function of position and let $\boldsymbol{u} \equiv (u_\varpi, u_\phi, u_z) \equiv u_\varpi \boldsymbol{e}_\varpi + u_\phi \boldsymbol{e}_\phi + u_z \boldsymbol{e}_z$ be a vector function of position. The standard operators of vector calculus take the following forms:

$$\nabla \psi = \left(\frac{\partial \psi}{\partial \varpi}, \frac{1}{\varpi} \frac{\partial \psi}{\partial \phi}, \frac{\partial \psi}{\partial z} \right) ; \tag{B.1}$$

$$\nabla \cdot \boldsymbol{u} = \frac{1}{\varpi} \frac{\partial}{\partial \varpi} (\varpi u_\varpi) + \frac{1}{\varpi} \frac{\partial u_\phi}{\partial \phi} + \frac{\partial u_z}{\partial z} , \tag{B.2}$$

$$\nabla \times \boldsymbol{u} = \frac{1}{\varpi} \begin{vmatrix} \boldsymbol{e}_\varpi & \varpi \boldsymbol{e}_\phi & \boldsymbol{e}_z \\ \partial/\partial \varpi & \partial/\partial \phi & \partial/\partial z \\ u_\varpi & \varpi u_\phi & u_z \end{vmatrix} . \tag{B.3}$$

Also

$$\nabla^2 \psi = \frac{1}{\varpi} \frac{\partial}{\partial \varpi} \left(\varpi \frac{\partial \psi}{\partial \varpi} \right) + \frac{1}{\varpi^2} \frac{\partial^2 \psi}{\partial \phi^2} + \frac{\partial^2 \psi}{\partial z^2} . \tag{B.4}$$

Note that the position vector is $\boldsymbol{r} \equiv \varpi \boldsymbol{e}_\varpi + z \boldsymbol{e}_z$ in cylindrical polar coordinates.

The only non-zero derivatives of the base unit vectors with respect to the coordinates are

$$\frac{\partial \boldsymbol{e}_\varpi}{\partial \phi} = \boldsymbol{e}_\phi , \qquad \frac{\partial \boldsymbol{e}_\phi}{\partial \phi} = - \boldsymbol{e}_\varpi .$$

B.2   Spherical Polar Coordinates (r, θ, ϕ)

In spherical polar coordinates (r, θ, ϕ), let ψ be a scalar function of position and let $\boldsymbol{u} \equiv (u_r, u_\theta, u_\phi) \equiv u_r \boldsymbol{e}_r + u_\theta \boldsymbol{e}_\theta + u_\phi \boldsymbol{e}_\phi$ be a vector function of position. The standard operators of vector calculus take the following forms:

$$\nabla\psi = \left(\frac{\partial\psi}{\partial r}, \frac{1}{r}\frac{\partial\psi}{\partial\theta}, \frac{1}{r\sin\theta}\frac{\partial\psi}{\partial\phi}\right) ; \tag{B.5}$$

$$\nabla\cdot\boldsymbol{u} = \frac{1}{r^2}\frac{\partial}{\partial r}(r^2 u_r) + \frac{1}{r\sin\theta}\frac{\partial}{\partial\theta}(\sin\theta\, u_\theta) + \frac{1}{r\sin\theta}\frac{\partial u_\phi}{\partial\phi} ; \tag{B.6}$$

$$\nabla\times\boldsymbol{u} = \frac{1}{r^2\sin\theta}\begin{vmatrix} \boldsymbol{e}_r & r\boldsymbol{e}_\theta & r\sin\theta\boldsymbol{e}_\phi \\ \partial/\partial r & \partial/\partial\theta & \partial/\partial\phi \\ u_r & ru_\theta & r\sin\theta u_\phi \end{vmatrix} . \tag{B.7}$$

Also

$$\nabla^2\psi = \frac{1}{r^2}\frac{\partial}{\partial r}\left(r^2\frac{\partial\psi}{\partial r}\right) + \frac{1}{r^2\sin\theta}\frac{\partial}{\partial\theta}\left(\sin\theta\frac{\partial\psi}{\partial\theta}\right) + \frac{1}{r^2\sin^2\theta}\frac{\partial^2\psi}{\partial\phi^2} . \tag{B.8}$$

Note that the position vector is $\boldsymbol{r} \equiv r\boldsymbol{e}_r$ in spherical polar coordinates.

The only non-zero derivatives of the base unit vectors with respect to the coordinates are

$$\frac{\partial\boldsymbol{e}_r}{\partial\theta} = \boldsymbol{e}_\theta , \qquad \frac{\partial\boldsymbol{e}_\theta}{\partial\theta} = -\boldsymbol{e}_r ,$$

$$\frac{\partial\boldsymbol{e}_r}{\partial\phi} = \sin\theta\boldsymbol{e}_\phi , \qquad \frac{\partial\boldsymbol{e}_\theta}{\partial\phi} = \cos\theta\boldsymbol{e}_\phi , \qquad \frac{\partial\boldsymbol{e}_\phi}{\partial\phi} = -\sin\theta\boldsymbol{e}_r - \cos\theta\boldsymbol{e}_\theta .$$

Appendix C

Self-adjoint Eigenvalue Problems

In considering stellar pulsation (Chapter 11) the eigensolutions of a self-adjoint differential operator were used. Here we summarize some of the properties of such self-adjoint eigenvalue problems. Suppose then that the differential equation

$$\mathcal{L}y = \lambda w y \qquad 0 < x < 1\,, \tag{C.1}$$

with suitable boundary conditions at $x = 0, 1$, has eigenvalues λ and eigenfunctions y (w being a prescribed positive function of x), and that the operator $\mathcal{L}$ is self-adjoint, i.e.

$$\int_0^1 y^* \mathcal{L}z \,\mathrm{d}x = \int_0^1 z(\mathcal{L}y)^* \,\mathrm{d}x \tag{C.2}$$

for arbitary y and z satisfying the boundary conditions. Then certain nice properties follow.

C.1 Reality of Eigenvalues

From Eq. (C.1) we have that $\lambda w y^* y = y^* \mathcal{L}y$ and hence

$$\lambda \int_0^1 w|y|^2 \mathrm{d}x = \int_0^1 y^* \mathcal{L}y \mathrm{d}x\,.$$

Taking the complex conjugate,

$$\lambda^* \int_0^1 w|y|^2 \mathrm{d}x = \int_0^1 y(\mathcal{L}y)^* \mathrm{d}x = \int_0^1 y^* \mathcal{L}y \mathrm{d}x\,,$$

where the last equality follows because $\mathcal{L}$ is self-adjoint. From this, comparing these two expressions, it follows that $\lambda = \lambda^*$; hence λ is real. Thus the eigenvalues are real.

C.2   Orthogonality of Eigenfunctions

Suppose

$$\mathcal{L}y_1 = \lambda_1 w y_1 \,, \qquad \mathcal{L}y_2 = \lambda_2 w y_2 \,. \qquad (C.3)$$

Taking the complex conjugate of the second equation and multiplying this by y_1, and subtracting this from the first equation multiplied by y_2^*, yields

$$(\lambda_1 - \lambda_2) w y_2^* y_1 \;=\; y_2^* \mathcal{L}y_1 - y_1(\mathcal{L}y_2)^* \,.$$

Integrating this gives

$$(\lambda_1 - \lambda_2) \int_0^1 w y_2^* y_1 \mathrm{d}x \;=\; \int_0^1 [y_2^* \mathcal{L}y_1 - y_1(\mathcal{L}y_2)^*] \mathrm{d}x \;=\; 0 \qquad (C.4)$$

because $\mathcal{L}$ is self-adjoint. Hence if $\lambda_1 \neq \lambda_2$,

$$\int_0^1 w y_2^* y_1 \mathrm{d}x \;=\; 0 \,, \qquad (C.5)$$

which is the orthogonality property. Thus, in this sense, the eigenfunctions corresponding to distinct eigenvalues are orthogonal.

C.3   Eigenfunction Expansions

We assume that the eigenfunctions are complete, so that (in an appropriate sense) any function $u(x)$ can be expressed as

$$u \;=\; \sum_{k=0}^{\infty} a_k y_k \qquad (C.6)$$

for some coefficients a_k. To determine a_n, multiply both sides of Eq. (C.6) by $w y_n^*$ and integrate to get (using orthogonality)

$$\int_0^1 w y_n^* u \mathrm{d}x \;=\; a_n \int_0^1 w|y_n|^2 \mathrm{d}x \,,$$

whence it follows that

$$a_n \;=\; \frac{\int_0^1 w y_n^* u \mathrm{d}x}{\int_0^1 w|y_n|^2 \mathrm{d}x} \,. \qquad (C.7)$$

Often the eigenfunctions are *normalized* such that $\int_0^1 w|y_n|^2 \mathrm{d}x = 1$.

C.4 Variational Principle

Let the eigenvalues be λ_k, with corresponding eigenfunctions y_k. We order the eigenvalues so that $\lambda_0 < \lambda_1 < \lambda_2 < \cdots$. From Eq. (C.1) we have that

$$\lambda_n = \frac{\int_0^1 w y_n^* \mathcal{L} y_n \, \mathrm{d}x}{\int_0^1 w|y_n|^2 \mathrm{d}x}. \tag{C.8}$$

Let us define

$$F[y] = \frac{\int_0^1 w y^* \mathcal{L} y \, \mathrm{d}x}{\int_0^1 w|y|^2 \mathrm{d}x} \tag{C.9}$$

for *any* function y satisfying the boundary conditions. Writing $y = \sum a_k y_k$ and using orthogonality (assuming the eigenfunctions to be normalized), it follows that

$$\int_0^1 w|y|^2 \mathrm{d}x = \sum_{k=1}^\infty |a_k|^2$$

and

$$\int_0^1 w y^* \mathcal{L} y \, \mathrm{d}x = \sum_{k=1}^\infty \lambda_k |a_k|^2. \tag{C.10}$$

Thus

$$F[y] = \frac{\sum \lambda_k |a_k|^2}{\sum |a_k|^2}. \tag{C.11}$$

Note first that if y is "almost equal to" a particular eigenfunction y_n, in the sense that the coefficient a_n is of order unity and all other coefficients are of order ϵ, a small quantity, then $F[y] = \lambda_n + O(\epsilon^2)$. Thus even if we have an estimate of the eigenfunction that is only accurate to first order in ϵ, we can get an estimate of the corresponding eigenvalue that is accurate to *second* order in ϵ. This is the variational principle. Note also from Eq. (C.11) that the minimum possible value of $F[y]$ is λ_0: thus for any y we choose, $F[y]$ provides an upper bound on the smallest eigenvalue.

Appendix D

The JWKB Method

The JWKB method (after Jeffreys, Wentzel, Kramers and Brillouin, though in fact it was used a century before any of them by Liouville and by Green), which we used in Chapter 12, provides an asymptotic description of waves in a slowly varying medium. We give only an outline of its derivation here.

Consider the wave-like equation

$$\frac{\mathrm{d}^2 y}{\mathrm{d}x^2} + K(x)y = 0 \, . \tag{D.1}$$

We assume that $y(x)$ is rapidly varying compared with $K(x)$. We seek a solution

$$y(x) = A(x)\exp\left[i\Psi(x)\right] \, , \tag{D.2}$$

where Ψ is a rapidly varying phase and A is a slowly varying amplitude function. Substituting (D.2) into Eq. (D.1) gives

$$\left(\frac{\mathrm{d}^2 A}{\mathrm{d}x^2} + 2i\frac{\mathrm{d}A}{\mathrm{d}x}\frac{\mathrm{d}\Psi}{\mathrm{d}x} + iA\frac{\mathrm{d}^2\Psi}{\mathrm{d}x^2} - A\left[\frac{\mathrm{d}\Psi}{\mathrm{d}x}\right]^2\right)\exp\left(i\Psi\right)$$
$$= -K(x)A(x)\exp\left(i\Psi\right) \, . \tag{D.3}$$

The dominant term on the left-hand side is the one containing $[\mathrm{d}\Psi/\mathrm{d}x]^2$, and we choose to make this balance the right-hand side:

$$\left[\frac{\mathrm{d}\Psi}{\mathrm{d}x}\right]^2 = K(x) \, , \qquad \text{hence} \quad \Psi = \int_{x_0}^{x} K^{1/2}\mathrm{d}x \, . \tag{D.4}$$

The next-order terms, linear in Ψ, must cancel:

$$2\frac{\mathrm{d}A}{\mathrm{d}x}\frac{\mathrm{d}\Psi}{\mathrm{d}x} + A\frac{\mathrm{d}^2\Psi}{\mathrm{d}x^2} = 0 \, , \qquad \text{hence} \quad A(x) = |K(x)|^{-1/4} \, . \tag{D.5}$$

This leaves the d^2A/dx^2 term in Eq. (D.3), which we neglect at this level of approximation since by assumption it is small compared with K.

Summarizing, we have

$$y(x) = \begin{cases} A\,|K(x)|^{-1/4}\cos\left(\int_{x_0}^x K(x')^{1/2}dx' + \phi\right) & \text{for } K(x) > 0 \\ A_\pm\,|K(x)|^{-1/4}\exp\left(\pm\int_{x_0}^x |K(x')|^{1/2}\,dx'\right) & \text{for } K(x) < 0 \end{cases} \tag{D.6}$$

for some suitable x_0, with A, $A_\pm$ and ϕ constants.

Note that the solution is oscillatory where $K(x) > 0$ and exponential where $K(x) < 0$. The approximation breaks down where K vanishes; hence we have to make a separate analysis in the neighbourhood of $K = 0$: this is also required to connect the solutions in the two regions $K > 0$ and $K < 0$.

Consider a turning point $x = x_1$ such that $K(x) < 0$ for $x < x_1$ and $K(x) > 0$ for $x > x_1$. Supposing that K has a simple zero, then close to x_1 we have

$$K(x) \approx (x - x_1)\,K_1 \tag{D.7}$$

where K_1 is a positive constant. Now Eq. (D.1) can be approximated as

$$\frac{d^2y}{dz^2} = -zy \tag{D.8}$$

where we have introduced the new variable

$$z = K_1^{1/3}(x - x_1)\,. \tag{D.9}$$

Equation (D.8) has general solution

$$y = C_1\text{Ai}(z) + C_2\text{Bi}(z)\,, \tag{D.10}$$

where C_1, C_2 are constants, and Ai and Bi are the Airy functions: see e.g. Abramowitz & Stegun (1964).

To be definite, we consider a solution that decreases exponentially in $x < x_1$. When $z < 0$ and $|z|$ is large, the Airy functions have asymptotic behaviour

$$\text{Ai}(z) \approx \frac{1}{2\sqrt{\pi}}|z|^{-1/4}\exp\left(-\frac{2}{3}|z|^{3/2}\right)\,,$$

$$\text{Bi}(z) \approx \frac{1}{2\sqrt{\pi}}|z|^{-1/4}\exp\left(\frac{2}{3}|z|^{3/2}\right)\,. \tag{D.11}$$

Thus we must have $C_2 = 0$ and so

$$y = C_1\,\text{Ai}(z) \tag{D.12}$$

in $x < x_1$. We can now use this to determine the phase ϕ in (D.6). For large positive z, the asymptotic behaviour of Ai is

$$\text{Ai}(z) \approx \frac{1}{\sqrt{\pi}} |z|^{-1/4} \cos \left(\frac{2}{3} |z|^{3/2} - \frac{\pi}{4} \right) . \qquad (D.13)$$

We assume that there is a region in which expressions (D.6) and (D.12) are both valid. Using approximation (D.7) for K, we have that

$$\Psi = \int_{x_1}^{x} K(x')^{1/2} dx' + \phi = \frac{2}{3} z^{3/2} + \phi \qquad (D.14)$$

and so Eq. (D.6) gives

$$y \approx A K_1^{-1/6} z^{-1/4} \cos \left(\frac{2}{3} z^{3/2} + \phi \right) . \qquad (D.15)$$

This agrees with Eq. (D.12) provided $\phi = -\pi/4$. Thus, sufficiently far from the turning point, the JWKB solution satisfying the boundary condition below $x = x_1$ is

$$y(x) = A_1 |K(x)|^{-1/4} \cos \left(\int_{x_1}^{x} K^{1/2}(x') \, dx' - \frac{\pi}{4} \right) . \qquad (D.16)$$

Similarly, if there is a second turning point $x_2 > x_1$ above which the solution must also be exponentially decaying, then the JWKB solution satisfying that boundary condition is

$$y(x) = A_2 |K(x)|^{-1/4} \cos \left(\int_{x}^{x_2} K^{1/2}(x') \, dx' - \frac{\pi}{4} \right) . \qquad (D.17)$$

To obtain a solution everywhere, we must be able to match these two solutions at a suitable point $x = x_f$ between x_1 and x_2. Define

$$\Psi_1 = \int_{x_1}^{x_f} K^{1/2}(x) \, dx - \frac{\pi}{4} , \quad \Psi_2 = \int_{x_f}^{x_2} K^{1/2}(x) \, dx - \frac{\pi}{4} . \qquad (D.18)$$

The condition that y and dy/dx are continuous at $x = x_f$ yields

$$A_1 \cos \Psi_1 = A_2 \cos \Psi_2 , \, -A_1 \sin \Psi_1 = A_2 \sin \Psi_2 . \qquad (D.19)$$

(The derivative of K has been neglected in this approximation.) The condition that there be a non-trivial solution for A_1 and A_2 is that the determinant of these linear equations vanishes:

$$\sin \Psi_1 \cos \Psi_2 + \cos \Psi_1 \sin \Psi_2 \equiv \sin(\Psi_1 + \Psi_2) = 0 , \qquad (D.20)$$

i.e. $\Psi_1 + \Psi_2 = (n-1)\pi$, where n is an integer. Thus we have the *quantization condition* that

$$\int_{x_1}^{x_2} K^{1/2}(x)\,dx = \left(n - \frac{1}{2}\right)\pi\,, \qquad n = 1, 2, \dots \,. \qquad \text{(D.21)}$$

Bibliography

Abramowitz, M. & Stegun, I. A., 1964. *Handbook of Mathematical Functions*, NBS Applied Mathematics Series No. 55 (NBS: Washington, DC).

Batchelor, G. K., 1953. *The Theory of Homogeneous Turbulence* (Cambridge University Press: Cambridge).

Batchelor, G. K., 1967. *An Introduction to Fluid Dynamics* (Cambridge University Press: Cambridge).

Beichman, C. A., 1987. *Ann. Rev. Astron. Astrophys.*, **25**, 521 – 563.

Benz, W., 1990. Smooth particle hydrodynamics: a review. In *The Numerical Modelling of Nonlinear Stellar Pulsations*, ed. J. R. Buchler (Kluwer: Dordrecht), p. 269 – 288.

Berger, M. J. & Oliger, J., 1984. Adaptive mesh refinement for hyperbolic-partial differential equations. *J. Comp. Phys.*, **53**, 484 – 512.

Bondi, H., 1952. On spherically symmetrical accretion. *Mon. Not. R. astr. Soc.*, **112**, 195 – 204.

Bouchy, F. & Carrier, F., 2001. P-mode observations on α Cen A. *Astron. Astrophys.*, **374**, L5 – L8.

Brandenburg, A., 2003. Computational aspects of astrophysical MHD and turbulence. In *Advances in Nonlinear Dynamos (The Fluid Dynamics of Astrophysics and Geophysics, Vol. 9)*, eds A. Ferriz-Mas & M. Núñez (Taylor & Francis: London), p. 269 – 344.

Brandenburg, A. & Dobler, W., 2002. Hydrodynamic turbulence in computer simulations *Comput. Phys. Commun.*, **147**, 471 – 475.

Chandrasekhar, S., 1969. *Ellipsoidal Figures of Equilibrium* (Yale University Press: New Haven, CT).

Chen, L. & Hasegawa, A., 1974. Plasma heating by spatial resonance of Alfvén wave. *Phys. Fluids*, **17**. 1399 – 1403.

Christensen-Dalsgaard, J., 2002. Helioseismology. *Rev. Mod. Phys.*, **74**, 1073 – 1129.

Christensen-Dalsgaard, J. & Berthomieu, G., 1991. Theory of solar oscillations. In *Solar Interior and Atmosphere*, eds A. N. Cox, W. C. Livingston & M. Matthews (University of Arizona Press: Tucson).

Clarke, C. J., Bonnell, I. A. & Hillenbrand, L. A., 2000. The formation of stellar clusters. In *Protostars and Planets IV*, eds V. Mannings, A. P. Boss & S. S. Russell (University of Arizona Press: Tucson, AZ), p. 151 – 177.

Cowling, T. G., 1934. The magnetic field in sunspots. *Mon. Not. R. astr. Soc.*, **94**, 39 – 48.

Cowling, T. G., 1976. *Magnetohydrodynamics* (Adam Hilger: Bristol).

Cox, J. P., 1980. *Theory of Stellar Pulsation* (Princeton University Press: Princeton, NJ).

Cushman-Roisin, B., 1994. *Introduction to Geophysical Fluid Dynamics* (Prentice Hall: Englewood Cliffs, NJ).

Deubner, F.-L. & Gough, D. O., 1984. Helioseismology: Oscillations as a diagnostic of the solar interior. *Ann. Rev. Astron. Astrophys.*, **22**, 593 – 619.

Drazin, P. G. & Reid, W. H., 1981. *Hydrodynamic Stability* (Cambridge University Press: Cambridge).

Duvall, T. L., Jr., 1982. A dispersion law for solar oscillations. *Nature*, **300**, 242 – 243.

Eardley, D. M. & Press, W. H., 1975. Astrophysical processes near black holes. *Ann. Rev. Astron. Astrophys.*, **13**, 381 – 422.

Elliott, J. R., 1996. *Solar Structure and Helioseismology*. Ph.D. thesis, Institute of Astronomy, University of Cambridge.

Fisher, G. H., Fan, Y., Longcope, D. W., Linton, M. G. & Pevtsov, A. A., 2000. The solar dynamo and emerging flux. *Solar Phys.*, **192**, 119 – 139.

Fryxell, B. *et al.*, 2000. FLASH: An adaptive mesh hydrodynamics code for modeling astrophysical thermonuclear flashes. *Astrophys. J. Suppl.*, **131**, 273 – 334.

Gill, A. E., 1982. *Atmosphere-Ocean Dynamics* (Academic Press: Orlando, FL).

Gough, D. O., 1969. The anelastic approximation for thermal convection. *J. Atmos. Sci.*, **26**, 448 – 456.

Gough, D. O., 1977. Random remarks on solar hydrodynamics. In *The Energy Balance and Hydrodynamics of the Solar Chromosphere and Corona*, eds R. M. Bonnet and P. Delache (G. de Bussac: Clermont-Ferrand), p. 3 – 36.

Gough, D. O., 1985. Inverting helioseismic data. *Solar Phys.*, **100**, 65 – 99.

Gough, D. O., 1993. Linear adiabatic stellar pulsation. In *Les Houches Session LXVII: Astrophysical Fluid Dynamics*, eds J.-P. Zahn & J. Zinn-Justin (Elsevier: Amsterdam), p. 399 – 560.

Hansen, C. J., 1978. Secular stability — applications to stellar structure and evolution. *Ann. Rev. Astron. Astrophys.*, **16**, 15 – 32.

Hernquist, L. & Katz, N., 1989. TREESPH — A unification of SPH wih the hierarchical tree method. *Astrophys. J. Suppl.*, **70**, 419 – 446.

Heyvaerts, J. & Priest, E. R., 1983. Coronal heating by phase-mixed Alfvén waves. *Astron. Astrophys.*, **117**, 220 – 234.

Holzer, T. E. & Axford, W. I., 1970. The theory of stellar winds and related flows. *Ann. Rev. Astron. Astrophys.*, **8**, 31 – 60.

Hoyle, F., 1953. On the fragmentation of gas clouds into galaxies and stars. *Astrophys. J.*, **118**, 513 – 528.

Ionson, J. A., 1978. Resonant absorption of Alfvénic surface waves and the heating of solar coronal loops. *Astrophys. J.*, **226**, 650 – 673.

Jeffreys, H. J. & Jeffreys, B., 1956. *Methods of Mathematical Physics* (3rd edition) (Cambridge University Press: Cambridge).

Kippenhahn, R. & Schlüter, A., 1957. Eine Theorie der solaren Filamente. *Zeitschrift für Astrophysik*, **43**, 36.

Kippenhahn, R. & Weigert, A., 1990. *Stellar Structure and Evolution* (Springer-Verlag: Berlin).

Kolmogorov, A. N., 1941. The local structure of turbulence in incompressible viscous fluid for very large Reynolds numbers. *Dokl. AN SSSR*, **30**, 299 – 303.

Lada, C. J. & Kylafis, N. D. (eds), 1991. *The Physics of Star Formation and Early Evolution* (Kluwer: Dordrecht).

Lada, C. J. & Lada, E. A., 2003. Embedded clusters in molecular clouds. *Ann. Rev. Astron. Astrophys.*, **41**, 57 – 115.

Landau, L. D. & Lifshitz, E. M., 1959. *Fluid Mechanics* (Pergamon Press: Oxford).

Larson, R. B., 1985. Cloud fragmentation and stellar masses. *Mon. Not. R. astr. Soc.*, **214**, 379 – 398.

Lebovitz, N. R., 1967. Rotating fluid masses. *Ann. Rev. Astron. Astrophys.*, **5**, 465 – 480.

Lighthill, J., 1978. *Waves in Fluids* (Cambridge University Press: Cambridge).

Lignières, F., 1999. The small-Péclet-number approximation in radiative interiors. *Astron. Astrophys.*, **348**, 933 – 939.

Lynden-Bell, D. & Ostriker, J. P., 1967. On the stability of differentially rotating bodies. *Mon. Not. R. astr. Soc.*, **136**, 293 – 310.

Lyttleton, R. A., 1953. *The Stability of Rotating Liquid Masses* (Cambridge University Press: Cambridge).

Mannings, V., Boss, A. P. & Russell, S. S. (eds), 2000. *Protostars and Planets IV* (University of Arizona Press: Tucson, AZ).

Mihalas D. & Mihalas, B. W., 1984. *Foundations of Radiation Hydrodynamics* (Oxford University Press: Oxford).

Moffatt, H. K., 1978. *Magnetic Field Generation in Electrically Conducting Fluids* (Cambridge University Press: Cambridge).

Monaghan, J. J. & Lattanzio, J. C., 1985. A refined particle method for astrophysical problems. *Astron. Astrophys.*, **149**, 135 – 143.

Monaghan, J. J., 1989. On the problem of penetration in particle methods. *J. Comput. Phys.*, **82**, 1 – 15.

Monaghan, J. J., 1992. Smoothed particle hydrodynamics. *Ann. Rev. Astron. Astrophys.*, **30**, 543 – 574.

Nelson, R. P. & Papaloizou, J. C. B., 1993. Three-dimensional hydrodynamic simulations of collapsing prolate clouds. *Mon. Not. R. astr. Soc.*, **265**, 905 – 920.

Niederreiter, H., 1978. Quasi-Monte Carlo methods and pseudo-random numbers. *Bull. Amer. Math. Soc.*, **84**, 957 – 1041.

Parker, E. N., 1955. Hydromagnetic dynamo models. *Astrophys. J.*, **122**, 293 – 314.

Parker, E. N., 1958. Dynamics of the interplanetary gas and magnetic field. *Astrophys. J.*, **128**, 664 – 676.

Parker, E. N., 1963. *Interplanetary Dynamical Processes* (Interscience: New York).

Pedlosky, J., 1979. *Geophysical Fluid Dynamics* (Springer-Verlag: New York).

Press, W. H., Flannery, B. P., Teukolsky, S. A. & Vetterling, W. T., 1986. *Numerical recipes* (Cambridge University Press: Cambridge).

Priest, E. R., 1993. Coronal heating mechanisms. In *Physics of Solar and Stellar Coronae*, eds J. Linsky and S. Serio (Kluwer: Dordrecht), p. 515 – 532.

Priest, E. R. & Forbes, T. G., 2000. *Magnetic Reconnection: MHD Theory and Applications* (Cambridge University Press: Cambridge).

Pringle, J. E., 1981. Accretion discs in astrophysics. *Ann. Rev. Astron. Astrophys.*, **19**, 137 – 162.

Proctor, M. R. E., 2003. Dynamo processes: the interaction of turbulence and magnetic fields. In *Stellar Astrophysical Fluid Dynamics*, eds M. J. Thompson & J. Christensen-Dalsgaard (Cambridge University Press: Cambridge), p. 143 – 158.

Proctor, M. R. E. & Gilbert, A. D. (eds), 1994. *Lectures on Solar and Planetary Dynamos* (Cambridge University Press: Cambridge).

Rayleigh, Lord, 1880. On the stability, or instability, of certain fluid motions. *Roy. London Math. Soc.*, **11**, 57 – 70.

Richtmyer R. D. & Morton, K. W., 1967. *Difference Methods for Initial-value Problems* (2nd edition) (Interscience: New York).

Roberts, B., 1981. Wave propagation in a magnetically structured atmosphere. I — Surface waves at a magnetic interface. *Solar Phys.*, **69**, 27 – 38.

Roberts, B., 1985. Magnetohydrodynamic waves. In *Solar System Magnetic Fields*, ed. E. R. Priest (Reidel: Dordrecht), p. 37 – 79.

Roberts, B., 1991. Magnetohydrodynamic waves in the Sun. In *Advances in Solar System Magnetohydrodynamics*, eds E. R. Priest and A. W. Hood (Cambridge University Press: Cambridge), p. 105 – 136.

Schiff, L. I., (1968). *Quantum Mechanics* (3rd edition) (McGraw-Hill: New York).

Shu, F. H., Adams, F. C. & Lizano, S., 1987. Star formation in molecular clouds: observation and theory. *Ann. Rev. Astron. Astrophys.*, **25**, 23 – 81.

Shu, F. H., 1992. *The Physics of Astrophysics Vol. II. Gas Dynamics* (University Science Books: Mill Valley, CA).

Snodgrass, H. B., 1983. Magnetic rotation of the solar photosphere. *Astrophys. J.*, **270**, 288 – 299.

Spiegel, E. A., 1971. Convection in stars. I. Basic Boussinesq convection. *Ann. Rev. Astron. Astrophys.*, **9**, 323 – 352.

Stone, J. M. & Norman, M. L., 1992. ZEUS-2D: A radiation magnetohydrodynamics code for astrophysical flows in two spatial dimensions. I — The hydrodynamic algorithms and tests. *Astrophys. J. Suppl.*, **80**, 753 – 790.

Tassoul, J.-L., 1978. *Theory of Rotating Stars* (Princeton University Press: Princeton).

Tassoul, M., 1980. Asymptotic approximations for stellar nonradial pulsations. *Astrophys. J. Suppl.*, **43**, 469 – 490.

Thompson, M. J., Christensen-Dalsgaard, J., Miesch, M. S. & Toomre, J., 2003. The internal rotation of the Sun. *Ann. Rev. Astron. Astrophys.*, **41**, 599 – 643.

Ulrich, R. K., Boyden, J. E., Webster, L., Padilla, S. P. & Snodgrass, H. B., 1988. Solar rotation measurements at Mount Wilson. V — Reanalysis of 21 years of data. *Solar Phys.*, **117**, 291 – 328.

Unno, W., Osaki, Y., Ando, H., Saio, H. & Shibahashi, H., 1989. *Nonradial Oscillations of Stars* (2nd edition) (University of Tokyo Press: Tokyo).

van Ballegooijen, A. A., 2000. Solar prominence models. In *Encyclopedia of Astronomy and Astrophysics*, ed. P. Murdin (Inst. of Physics: London).

Zahn, J.-P., 1993. Instabilities and turbulence in rotating stars. In *Les Houches Session LXVII: Astrophysical Fluid Dynamics*, eds J.-P. Zahn & J. Zinn-Justin (Elsevier: Amsterdam), p. 561 – 615.

Zel'dovich, Ya. B. & Raizer, Yu. P., 2002. *Physics of Shock Waves and High-Temperature Hydrodynamic Phenomena* (Dover Publications: New York).

Ziegler, U., 1998. NIRVANA$^+$: An adaptive mesh refinement code for gas dynamics and MHD. *Comput. Phys. Commun.*, **109**, 111 – 134.

Index